T0135781

Technische Universität München
Fachgebiet Höchstfrequenztechnik

Evaluation of wide-beam, short-range synthetic aperture radar imaging

Florian Gerbl

Vollständiger Abdruck der von der Fakultät für Elektrotechnik und Informationstechnik der Technischen Universität München zur Erlangung des akademischen Grades eines

Doktor-Ingenieurs

genehmigten Dissertation.

Vorsitzender:	Univ.-Prof. Dr. sc. techn. (ETH) Andreas Herkersdorf
Prüfer der Dissertation:	1. Univ.-Prof. Dr.-Ing., Dr.-Ing. habil. Erwin Biebl
	2. Univ.-Prof. Dr.-Ing., Dr.-Ing. habil. Robert Weigel, Friedrich-Alexander-Universität Erlangen-Nürnberg

Die Dissertation wurde am 01.10.2007 bei der Technischen Universität München eingereicht und durch die Fakultät für Elektrotechnik und Informationstechnik am 30.11.2007 angenommen.

Bibliografische Information der Deutschen Nationalbibliothek

Die Deutsche Nationalbibliothek verzeichnet diese Publikation in der
Deutschen Nationalbibliografie; detaillierte bibliografische Daten sind
im Internet über http://dnb.d-nb.de abrufbar.

ISBN 978-3-8325-1831-8

Logos Verlag Berlin GmbH
Comeniushof, Gubener Str. 47,
10243 Berlin
Tel.: +49 030 42 85 10 90
Fax: +49 030 42 85 10 92
INTERNET: http://www.logos-verlag.de

Danksagung

Die vorliegende Arbeit entstand im Rahmen meiner Tätigkeit am Fachgebiet Höchstfrequenztechnik der Technischen Universität München.

Mein herzlicher Dank gilt Herrn Prof. Dr. Erwin Biebl, der mit seiner fachlichen Kompetenz und seiner freundlichen Art ideale Bedingungen für die Durchführung dieser Arbeit geschaffen hat.

Den Kollegen Dr.-Ing. Mark Beer, Andreas Fackelmeier, M. Sc., Dipl.-Ing. Stefan Holzknecht, Dr.-Ing. Anton Lindner, Dipl.-Ing. Christian Morhart, Dipl.-Ing. Magnus Olbrich, Dipl.-Ing. Florian Pfeiffer, Dr.-Ing. Florian Ramian und Dr. rer. nat. Michael Streifinger möchte ich für die zahlreichen fachlichen Diskussionen und das äußerst angenehme Arbeitsklima, sowie Yu Lin, M. Sc., und Ing. (ENSEIRB) Daniel Mineau für die Entwicklung von Systemkomponenten danken.

Den Herren Manfred Agerer, Josef Franzisi und Hans Mulatz danke ich für die rasche und präzise Durchführung feinmechanischer Arbeiten.

Ganz besonders herzlich danke ich meinen Eltern und meiner Frau Dagmar für ihre stets liebevolle Unterstützung.

Contents

Contents

List of Figures

1 Introduction

It is one of the characteristics of today's life that an increasing number of tasks, which so far have been performed by humans, is either assisted or completely taken over by intelligent machines. To fulfill their tasks, machines have to rely on sensors that provide them with information on their environment. There are sensors susceptible to physical phenomena for which also humans have a sense—e.g. charge coupled devices (CCD), which are susceptible to light as eyes are—and there are sensors that are capable of exploiting physical phenomena which are not accessible to human sense organs—like electromagnetic waves outside the visible region, e.g. radio signals. Belonging to the latter class are radio detection and ranging (radar) systems. Historically, they have been applied in sectors where high costs were affordable, as e.g. in military and aviation. Nowadays, radar systems are entering the mass market as they are e.g. included in upper-class vehicles in order to assist the driver by providing automatic cruise control functionality. For such a system to be economically successful, it has to be producible very cost-effectively. The keys to a cost-effective realization are a thorough understanding of the underlying principles and the ability to decide which features a system really needs to provide, in order to accomplish the task it is expected to. In this work, such properties—amongst others—of short-range radar imaging will be discussed that might allow extremely cost-effective implementations for certain imaging tasks. Potential applications of short-range radar imaging systems include such of the automotive sector (e.g. as a supplement to ultrasound parking aids), the agricultural sector (e.g. for the detection and rescue of wild animals [1]), and industry (e.g. for automation purposes), to name a few.

1.1 State of the art

The term *radar* describes a class of techniques to detect and localize objects that interact with electromagnetic waves. A century ago, Christian Hülsmeyer was the first one to build a system—called "Telemobiloskop"—that was able to detect the presence of an object—in his case a ship—by means of electromagnetic waves [2]. An overview of radar's evolution from then to today can be found in [3].

A key characteristic of radar systems is their ability to recognize that two closely spaced objects—commonly called *targets* in radar terminology—are actually two objects and not a single one. This property is called *resolution*. Radial resolution is commonly obtained by transmitting signals exhibiting a certain bandwidth, where higher bandwidth yields better resolution. In systems with *real apertures*, which were the first systems to be developed, lateral resolution is determined by the antenna, used for transmitting and receiving, and the distance to the target. Pointing the antenna to different directions, targets can be determined to lie within or outside the illuminated region. The smaller the extent of the antenna beam is at the distance of the target, the smaller is the allowable lateral spacing between two targets so that they still can be resolved.

In *synthetic aperture radar* (SAR) systems, lateral resolution is achieved by moving the system along the scene that is to be imaged and coherently processing the signals that are received along the system's path. The term *SAR imaging* stands for a variety of techniques rather than a single one. The first SAR system, called "Doppler beam sharpener," was developed by Wiley [4] in the early 1950s. He recognized that the signals reflected from different targets as an airborne radar system flies above a scene exhibit different Doppler shifts depending on the targets' positions, and that this information can be used to locate the targets regarding their positions along the flight path. Spaceborne SAR systems have been developed for imaging and monitoring the Earth from satellites. There are SAR systems operating in free-space and such that are foliage- or ground-penetrating, used e.g. for landmine detection or non-destructive testing [5]. Ground- and foliage-penetrating radars commonly operate at frequencies below and around 1 GHz, which are preferable over higher frequencies with their associated higher attenuation [6]. This work will be focused on free-space, ground-based, low-cost, short-range SAR imaging using frequencies around 24 GHz, where *short-range* means a few centimeters to meters. SAR systems for ranges between a few centimeters and a few hundred meters have been reported e.g. in [7–22]. Some of them are used to detect changes in the environment by interferometric processing. Due to the limited lengths of their synthetic apertures as compared to the targets' distances, those systems cover only narrow regions of target aspect angles. Systems for shorter ranges, many of which are used for the detection of buried objects, commonly operate at frequencies well below 24 GHz [7, 11, 14, 15, 17–19, 21] and/or use expensive vector network analyzers as key components [8, 9, 12, 13, 16–18, 20, 22]. In [10], a system is described that operates around 10 GHz and that is claimed to be a low-cost system. Its radiofrequency hardware comprises two signal sources that cost a few hundred US Dollars each.

Horn antennas are widely-used as transmitting and receiving elements of short- and medium-range SAR systems [7, 10–13, 15, 20, 22]. Their comparatively narrow antenna beams, in conjunction with matched filters that are widely-used as part of the signal processing, limit the region of aspect angles, from which targets are effectively observed. That eases sampling requirements and potentially allows approximations in the processing algorithms, but consideration of wider regions of aspect angles might yield interesting image properties.

1.2 Goal of this work

It is the intent of this work to present the reader properties of wide-beam, short-range synthetic aperture radar imaging, that might be used advantageously for implementing cost-effective systems and complying with official frequency regulations, and to present a radar system developed for validating aspects that will be treated theoretically.

From the many standpoints, that SAR processing can be treated from, the one of *mean reflectivity estimation*—where the mean is with respect to the aspect angles from which the targets are observed by the system—is chosen to provide a framework, in which different imaging tasks can be unified.

Special attention is paid to amplitude variations of the signals that are processed— an issue commonly disregarded in the literature but inevitable for a concise treatment of wide antenna beams. Additionally, including amplitude variations allows to exactly understand the influence of applying windowing functions not only on image properties, like resolution and side-lobe levels, but also on the potential result of unequal pixel values for targets with equal reflectivity characteristics but different orientation. Simulation results will illustrate the analytical derivations.

The beam-width of the transmit and receive antennas has great influence on the information that is conveyed in a SAR image. A figure of merit will be introduced that allows to judge the fidelity of the SAR image in dependence of target characteristics and the antenna's beam-width. For a variety of imaging tasks, antenna patterns will be derived that maximize the fidelity, and the influence of using suboptimal patterns will be discussed.

Realization issues will be addressed, and a system implemented for validating imaging concepts will be reported on. Imaging experiments showing the benefits of wide-beam processing and the capabilities of low-cost implementations, amongst others, will round off the work.

1 Introduction

2 Issues on wide-beam, short-range synthetic aperture radar imaging

Radar systems achieve information on the radar reflectivity distribution of their environment by transmitting electromagnetic signals and receiving and processing the signals echoed by their surrounding. Depending on the bandwidth of the transmitted signal, targets that are located at different *ranges*—as the distances to the targets are commonly called—can be resolved. For fixed antenna position and orientation, the only possible statement on the direction of a detected target is, that it must be located somewhere inside the region that is illuminated by the system's antenna. The angular extent of the illuminated region depends on the size of the antenna's aperture. As the size of the aperture increases with respect to the wavelength of the transmitted signal, the angular extent of the illuminated region decreases. The reflectivity information provided by such a system is one-dimensional for the direction into which the antenna is pointed. Rotating the antenna, as done e.g. in airport surveillance radar systems, and displaying the one-dimensional reflectivity information obtained for each orientation versus the antenna's angle of rotation yields a two-dimensional reflectivity map, where the angular resolution depends on beam-width and therefore on the size of the antenna's aperture. Instead of rotating the antenna at a fixed location, it can be moved along a straight path and directed into a fixed direction with respect to that path, as it is done e.g. in so-called side-looking airborne radar (SLAR) systems. Again, the one-dimensional reflectivity information with respect to range obtained at each of the positions along the path can be displayed versus the position, resulting in a two-dimensional reflectivity map. The resolution in the direction of the system's motion—called *lateral* resolution for the moment—is determined not only by the angular resolution—given by the antenna's aperture size and the wavelength—but also on the range of the target, since the distance, that the radar system has to pass until a target leaves the illuminated region, depends on the target's distance from the system's line of motion. For a given desired lateral resolution in great distances from the radar system, the required antenna aperture might be prohibitively large—potentially on the order of some hundred meters for spaceborne systems. Related to the principle of Doppler beam sharpening [4], methods have been developed that emulate large antenna apertures by properly processing the data obtained using antennas with small *real* apertures along a defined path—the so called *synthetic* aperture. In contrast to systems with real apertures, the lateral resolution does not depend on the range of the target. However, for a given lateral resolution, a certain angular extent has to be covered by the synthetic aperture, which means that the necessary length of the synthetic aperture increases with increasing range for given lateral resolution.

There are a variety of points of view from which to look at synthetic aperture radar imaging. Among those mentioned in [23] are: synthesized antenna aperture, Doppler beam sharpening, correlation with reference point-target response, matched filter for received point-target signal, de-chirping of Doppler frequency shift, and optical-focusing equivalent, each of which is covered in the literature. Commonly considered the most prevalent task of SAR systems is the generation of high-resolution reflectivity maps—the SAR images. Throughout SAR literature, it is mentioned that the magnitudes displayed in SAR images are estimates of the weighted averages of the targets' reflectivities for different aspect angles, but little attention is paid to the weighting in effect due to the processing actually implemented. Matched filters—or approximations that speed-up the processing—are commonly used to generate SAR images. As a consequence, the weighting of a target's reflectivity contributions from different aspect angles is non-uniform in general. That means, that the magnitude of that pixel of the SAR image that corresponds to the location of the target may depend on the target's orientation. As an example, consider an idealized target that exhibits zero reflectivity except for a certain direction. In case of non-uniform reflectivity weighting, the magnitude of the target's pixel is given by the weighting introduced by the algorithm for that particular direction, for which the target reflects the incident wave to the radar system. It might be undesirable to obtain varying magnitudes—and in consequence varying signal-to-noise ratios—for equal targets that differ only by their orientation. Not only for targets of that class, assigning equal weights to reflectivity contributions from all observed angles might be desirable.

2.1 Imaging setup and coordinate systems

A number of modes of SAR imaging exist, differing in the trajectory along which data are recorded and in whether the direction of the antenna with respect to the trajectory of the system is kept constant. The discussion will be focused on a *linear* SAR in *stripmap* mode, which means that the trajectory of the system is linear and the orientation of the antenna with respect to the system's trajectory is fixed. Additionally, the antenna is assumed to be *unsquinted*, which means that the resulting illumination of the scene is symmetric about a plane perpendicular to the system's path. The imaging setup considered throughout this text is illustrated in fig. 2.1. The radar system, indicated by plots of an arbitrarily chosen antenna pattern at different locations, is moved linearly along the scene to be imaged, which is represented by the plot of an arbitrarily chosen reflectivity characteristic of a single target with unknown orientation. Equidistantly along the system's path, a signal is transmitted, reflected by the scene according to the reflectivity of the region that is illuminated by the antenna beam, and received and stored by the system, resulting in discrete data to be processed rather than in continuous data. The processing, however, will be discussed in terms of integrals rather than discrete sums whenever feasible, as is common in SAR literature, since that simplifies the treatment. The antennas of transmitter and receiver are assumed to be co-located, resulting in a configuration commonly called *monostatic*.

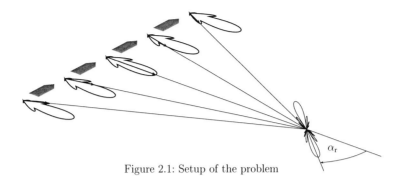

Figure 2.1: Setup of the problem

As it will become clear from the subsequent sections, for the generation of a SAR image, only the relative motion of the radar system with respect to the target is essential, not any absolute motion. Therefore, the motion can be interpreted as a motion of the radar system in a coordinate system fixed to the scene, and equivalently as a motion of the scene in a coordinate system fixed to the radar. That will allow to describe a given problem from either point of view, potentially simplifying the discussion.

Three coordinate systems, depicted in fig. 2.2, will be used to describe the locations and motions of the targets and the antenna: the scene-centered Cartesian (x, y, z)-coordinate system, the antenna-centered Cartesian (u, v, w)-coordinate system, and the antenna-centered spherical (r, α, β)-coordinate system. The (u, v, w)-coordinate system is a copy of the (x, y, z)-coordinate system, shifted by x_a in x-direction.

A target located at $(x_\mathrm{t}, y_\mathrm{t}, z_\mathrm{t})$ in the (x, y, z)-coordinate system and thus at $(x_\mathrm{t} - x_\mathrm{a}, y_\mathrm{t}, z_\mathrm{t})$ in the (u, v, w)-coordinate system has a distance $r_0 = \sqrt{y_\mathrm{t}^2 + z_\mathrm{t}^2}$ to both the x- and the u-axis. As it moves by the antenna, its distance to the antenna is

$$r = \sqrt{u_\mathrm{t}^2 + y_\mathrm{t}^2 + z_\mathrm{t}^2} = \sqrt{u_\mathrm{t}^2 + r_0^2}, \tag{2.1}$$

or

$$r = \sqrt{(x_\mathrm{t} - x_\mathrm{a})^2 + r_0^2} \tag{2.2}$$

in terms of $x_\mathrm{t} = u_\mathrm{t} + x_\mathrm{a}$.

The variation of r as the system passes the target, and the associated phase shift between the transmitted and the received signals, will turn out to be the key features in forming a SAR image. The dependence of r on r_0 rather than on the particular values of y_t^2 and z_t^2 results in a cylinder symmetry, which means that targets with equal x-positions and equal r_0—irrespective of the actual combination of y_t and z_t for a given r_0—cannot be resolved, since the phase shifts observed along the synthetic aperture will be equal.

With the investigation of antenna patterns, it will prove convenient to describe the direction to the target, as observed by the radar system, in a spherical coordinate

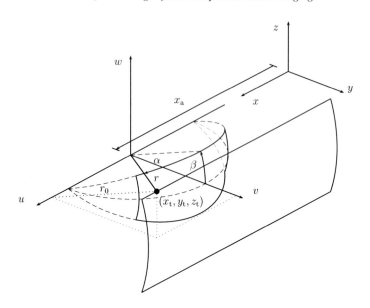

Figure 2.2: Antenna-centered spherical coordinate system

system. In SAR literature, different terms are used to refer to different directions in the imaging geometry. The direction perpendicular to the path of the radar system is commonly called *range*, *cross-track* or *fast-time* direction, while the direction along the path is called *cross-range*, *along-track*, *slow-time* or *azimuth* direction. This fact might suggest to orient a spherical coordinate system such that its *azimuth* angle is measured from the u-axis in the u-v-plane. However, rotating the spherical coordinate system by $90°$ about the v-axis offers several advantages that will be discussed next.

The antenna-centered Cartesian and the spherical coordinate systems in fig. 2.2 are related by

$$u = r \sin \alpha \tag{2.3}$$
$$v = r \cos \alpha \cos \beta \tag{2.4}$$
$$w = r \cos \alpha \sin \beta, \tag{2.5}$$

which means that

$$\alpha = \arcsin \left(\frac{u}{r} \right) \tag{2.6}$$

and

$$\beta = \arctan \left(\frac{w}{v} \right). \tag{2.7}$$

Although the target is moving in x-direction, β—since it is independent of r—is constant according to (2.7). That means that the direction to the target as it passes by the radar system—with y and z and therefore v and w constant—can be described by one single varying angle coordinate, α, and a fixed angle coordinate, β, resulting in concise and neat formulations of antenna patterns in subsequent sections. Limiting an antenna's radiation to $\pm\alpha_0$ and $\pm\beta_0$ according to the chosen orientation results in a field of view, whose extent in x-direction, for a given r_0, does not depend on β. That means, that the lengths of the synthetic apertures for targets that are not resolvable anyway are equal. Conversely, equal aperture lengths for such unresolvable targets can be forced by limiting the radiation to a certain range of α, independently of β.

2.2 The role of the temporal Doppler effect

It is well known that the formation of a SAR image relies on the Doppler effect. The first implementation of what is considered a SAR system today was called a "Doppler beam sharpener" [4], hinting to the underlying principle. However, not the temporal Doppler effect due to the relative *velocity* between radar and target is crucial in forming the image—actually it might have to be corrected for)—but the change in *range* between radar system and target as they pass by each others. This change of *range* results in a characteristic phase signature along the synthetic aperture, which is the property that is exploited to form the SAR image. To determine whether or not the temporal Doppler shift has to be corrected for, it will be quantified in terms of parameters of the imaging system.

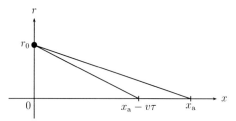

Figure 2.3: Geometry for the investigation of the temporal Doppler effect

Consider the setup in fig. 2.3. The radar system transmits an electromagnetic wave, whose type—e.g. a pulse or a continuous-wave (CW) signal—depends on the radar system, at a certain position and receives the backscattered wave at a slightly different position, since the system is moving while the wave propagates to the target and back. Thus, the phase information corresponding to the range at the transmit position, which is the information that is essential in forming the SAR image, is corrupted by the motion of the radar system.

9

The phase information is contained in the round-trip time of the transmitted pulse. For a radar system at the (x, y, z) coordinates $(x_a, 0, 0)$ and a target at the position $(0, r_0 \cos \beta_t, r_0 \sin \beta_t)$, where the target's x-coordinate has been chosen zero without loss of generality, the range r is

$$r(x_a) = \sqrt{x_a^2 + r_0^2}. \tag{2.8}$$

The corresponding round-trip time is

$$\tau(x_a) = \frac{2r(x_a)}{c}, \tag{2.9}$$

where c is the propagation speed of the wave, and the associated phase of the received signal with respect to the transmitted signal is

$$\phi(x_a) = \omega \tau(x_a), \tag{2.10}$$

in case the target itself does not introduce an additional phase-shift, where $\omega = 2\pi f$ is the frequency of the transmitted signal.

Synthesizing the aperture in a stop-and-go manner—i.e. taking a certain position, transmitting a signal and recording the echo, moving the system to the next position, stopping, and performing the next measurement—the desired phase signature can be recorded without errors originating from a movement of the radar system. However, the motion of the radar system while transmitting and receiving causes the round-trip time to change. Assuming the radar system moving at a constant velocity v along the x-axis, the signal received at time t at the position $x_a = vt$ is that signal that has been transmitted a time span τ earlier at the position $x_a - v\tau$. According to fig. 2.3, the time between transmission at $x_a - v\tau$ and reception at x_a, τ, is given by the implicit expression

$$\tau = \frac{1}{c} \left[\sqrt{(x_a - v\tau)^2 + r_0^2} + \sqrt{x_a^2 + r_0^2} \right]. \tag{2.11}$$

A similar expression is given in [24, p. 158 f], but the discussion there is truncated with the statement that v is much lower than c and therefore the Doppler effect does not have to be considered. The same conclusion will be drawn from the following calculations, however, the influence of the Doppler effect will be quantified, and it will turn out that the fact that the system's velocity is much lower than the velocity of light alone is not a sufficient justification for neglecting the Doppler effect.

Solving the above implicit expression for τ yields

$$\tau = 2 \frac{c\sqrt{x_a^2 + r_0^2} - x_a v}{c^2 - v^2}. \tag{2.12}$$

The strength of the Doppler effect depends on the radial velocity between system and target, which for a given velocity v increases with increasing aspect angle α. Therefore, the discussion can be focused to large aspect angles, which means that r_0 can be neglected in (2.12) and thus τ reduces to

$$\tau = \frac{2x_a}{c + v}. \tag{2.13}$$

The phase according to the path length from the transmitter to the target and back to the receiver is measured by mixing the transmitted harmonic signal

$$\cos(2\pi f t) \tag{2.14}$$

with the received signal and low-pass filtering the mixer output to suppress signal components around $2f$. Assuming that the travel times from the transmitter to the target and that from the target to the receiver are approximately equal, the signal at the target at time t is the one that has been transmitted a time span $\tau/2$ earlier, and therefore reads as

$$\cos\left(2\pi f \left(t - \frac{\tau}{2}\right)\right), \tag{2.15}$$

and with $x_a = vt$ as

$$\cos\left(2\pi f t \left(1 - \frac{v}{c+v}\right)\right). \tag{2.16}$$

The signal at the receiver at time t is the one reflected by the target a time span $\tau/2$ earlier, and therefore reads as

$$\cos\left(2\pi f t \left(1 - \frac{2v}{c+v} + \frac{v^2}{(c+v)^2}\right)\right), \tag{2.17}$$

and for $v \ll c$ approximately as

$$\cos\left(2\pi f t \left(1 - \frac{2v}{c} + \frac{v^2}{c^2}\right)\right). \tag{2.18}$$

Mixing with the transmitted signal given by (2.14) and low-pass filtering yields the phase difference between transmitted and received signal to be

$$\phi = 2k x_a + k x_a \frac{v}{c}, \tag{2.19}$$

where $2k x_a$ is the desired term containing distance information, and $k x_a v/c$ is an undesired term. Exactly the same result is obtained by assuming the received signal to be shifted in frequency—with respect to the signal it is mixed with—by the Doppler frequency [25, p.97ff]

$$f_D = -\frac{2v}{c} f, \tag{2.20}$$

where the negative sign results from the fact that the Doppler frequency is negative for increasing x_a. That means that the undesired term in (2.19),

$$\Delta\phi_D = k x_a \frac{v}{c} = \frac{2\pi f x_a v}{c^2}, \tag{2.21}$$

quantifies the influence of the Doppler effect. It is obvious, that the phase error not only depends on the ratio v/c, but also on the distance to the target. The maximum allowable phase error depends on the actual application. For a hypothetic maximum error $|\Delta\phi_{\rm D}| = \pi/8$, a velocity $v = 15\,{\rm m/s}$, and a frequency $f = 24\,{\rm GHz}$, the maximum allowable target distance in x-direction is $x_{\rm a} = 8\,{\rm km}$. Therefore, the temporal Doppler effect does not have to be compensated for in a short-range system at the given frequency. The same may be true for long-range systems, where comparatively narrow antenna beams limit the radial velocity to values much smaller than the system's velocity.

2.3 Signal model

In the following sections, different paradigms for the reconstruction of the reflectivity distribution of the scene will be discussed. Therefore, a notation will be chosen that allows for the evaluation of the influence of weighting functions on the meaning and the properties of the resulting SAR image. For this purpose, the signal observable along the synthetic aperture due the presence of a point-like target will be derived first.

The power $P_{\rm RX}$ received from a target with radar cross section σ at range r in a direction where the antennas directive gain is G, in response to an illumination by the radar system with an electromagnetic wave of power $P_{\rm TX}$ at the wavelength λ, is given by the *radar equation*

$$P_{\rm RX} = \frac{\lambda^2 \cdot P_{\rm TX} \cdot G^2 \cdot \sigma}{(4\pi)^3 \cdot r^4}, \tag{2.22}$$

which can be found in various similar forms throughout the literature, e.g. in [26].

Equation (2.22) does not convey information on the phase of the received signal. In pulse radars, power detection alone might be sufficient to determine the range of a target, and the received signal might be modeled sufficiently detailed by (2.22). For the formation of SAR images however, the system needs to be coherent, i.e. it has to be able to provide information on the phase difference between the transmitted and the received signal. Therefore, the received signal will be modeled more detailed with respect to what happens to the signal's phase on the way from the transmitter to the target and back.

The magnitude of the received signal will be taken as proportional to the square root of the received power, thus proportional to

$$u'_{\rm RX} = \frac{\lambda \cdot \sqrt{P_{\rm TX}} \cdot G \cdot \sqrt{\sigma}}{(4\pi)^{3/2} \cdot r^2}. \tag{2.23}$$

Additionally, phase terms accounting for various effects will be included. Due to the length of the path from the transmitter to the target and back to the receiver, $2r$, the received signal lags the transmitted signal by a phase difference $2kr$, where $k = 2\pi/\lambda$ is the wavenumber. A potential phase shift at the target is accounted for by replacing the real-valued term $\sqrt{\sigma}$ by the potentially complex-valued reflexion coefficient ρ, which are thus related by

$$|\rho|^2 = \sigma. \tag{2.24}$$

It is important to note that the reflexion coefficient depends on the polarization of the incident wave. Since it is dealt with scalar waves in this investigation rather than with vector waves, ρ is to be understood as the reflexion coefficient that relates the field components at the receiver in the polarization direction for which the receiver is sensitive, to the fields incident on the target according to the polarization of the transmitter. In general, the reflectivity of a target varies with the aspect angle, which is accounted for by letting ρ be a function of the aspect angle.

The antenna is another potential source of phase shifts. The antenna gain is generally a function of the direction. It relates the power density in a certain direction due to the given antenna to the power density produced by a lossless isotropic radiator in the same direction. The antenna gain G and the antenna directivity D are related by

$$G = \epsilon D, \tag{2.25}$$

where ϵ is the efficiency of the antenna [27]. The efficiency will be absorbed in a system constant in the following, and thus only the directivity will occur explicitly in the following equations. Note that the directivity is to be understood as a direction dependent function and not as the maximum of this function, as it is often the case in radar literature. Like with the (real-valued power) radar cross section σ and the (complex-valued amplitude) reflexion coefficient ρ, there is a complex-valued counterpart D_A to the antenna (power) directivity D, related by

$$|D_\mathrm{A}|^2 = D, \tag{2.26}$$

which will be referred to as the antenna *amplitude pattern*, indicated by the subscript A. The fact that D_A is complex-valued can be interpreted as a varying phase center of the antenna [27, p. 799 ff]. The terms *antenna directivity* and *antenna pattern* will be used interchangeably, potentially augmented by the terms *amplitude* or *power* where necessary. In order to account for the phase influence of the antenna, G in (2.23) has to be replaced by the complex-valued two-way amplitude pattern D_A^2 rather than by the real-valued power pattern D, which has the same magnitude as D_A^2 but does not convey phase information.

Incorporating the mentioned phase terms, the signal received along the synthetic aperture at antenna position x due to a point-like uniform reflector at $x = 0$ after coherent down-conversion, which will be referred to as the *reference* signal, can be modeled as

$$s_x(x) = \frac{a \cdot D_\mathrm{A}^2(\alpha(x)) \cdot \rho(\alpha(x)) \cdot \exp\{-\mathrm{j}2kr_x(x)\}}{r_x^2(x)}, \tag{2.27}$$

which means that the voltages at the inphase and the quadrature outputs of the receiver are proportional to the real and imaginary parts, respectively, of the (dimensionless) signal s_x, where the constant of proportionality is u_0, a voltage chosen arbitrarily in order to make s_x dimensionless. The antenna efficiency has been absorbed in the system constant a,

$$a = \sqrt{\frac{R \cdot P_\mathrm{TX} \cdot g}{(4\pi)^3}} \cdot \epsilon \cdot \lambda \cdot \frac{1}{u_0}, \tag{2.28}$$

as have been the wavelength, the transmitted power, mixer losses, mismatches and the like, merged in the factor g, which is potentially smaller than 1. R is the value of the load resistor, at which the signal voltage is measured. The subscripts x in (2.27), where used, indicate that the respective function is a function of x. Since e.g. the range r will be used in different domains and therefore with different arguments in subsequent sections, functions with equal meanings but different arguments will be supplied with the respective subscripts. As an example, consider the range r, which as a function of x reads as

$$r_x(x) = \sqrt{x^2 + r_0^2}, \tag{2.29}$$

whereas it reads as

$$r_\alpha(\alpha) = r_0 \cos \alpha \tag{2.30}$$

as a function of the angle α.

2.4 Solving the inverse problem—uniform reflectivity weighting

Equation (2.27) describes what is called the *forward problem*—the signal observable when a point-like target is present. It is the task of a SAR system to solve the *inverse problem*, i.e. to reconstruct the reflectivity distribution from the observed signal. In case the scene is described by a reflectivity *distribution* rather than a single reflectivity *pulse* corresponding to a point-like target, the signal observable is the convolution of the reference function and the reflectivity function. Let γ be the reflectivity distribution, where for the moment it is assumed that the reflectivity is not a function of the aspect angle, then the observed signal o is

$$o = \gamma * s_x. \tag{2.31}$$

According to the convolution theorem, this relation can be written as

$$O = \Gamma \cdot S_x \tag{2.32}$$

in the Fourier domain. That means that the reflectivity distribution can be reconstructed as

$$\gamma = \mathscr{F}^{-1} \left\{ \frac{O}{S_x} \right\} \tag{2.33}$$

from the observed signal, where $\mathscr{F}^{-1}$ denotes the inverse Fourier transform. This process is called *deconvolution*. A direct implementation of this deconvolution is not feasible in practice, since from measurements within a finite aperture, the spectrum of the observed signal has finite support, as will be discussed in section 2.5.1. A way to circumvent the associated problems is to define $1/S_x \equiv 0$ outside the support region of S_x, which is called the principal solution of such problems, see e.g. [24, p. 149]. However, this implementation is considered problematic as it is susceptible to additive noise in many applications, where the strength of the observed signal decreases at the edges of the (spatial) frequency band, and therefore noise contributions are amplified according to the inverse of the comparatively small values of S_x at the band edges. A common workaround for the noise problem is the use of a matched filter, see e.g. [28, p. 43], i.e. an estimate for the reflectivity distribution is obtained as

$$\gamma = \mathscr{F}^{-1}\left\{O \cdot S_x^*\right\}, \tag{2.34}$$

which in the spatial domain is the correlation of the observed signal and the signal expected for a point-target, i.e. the reference function. Instead of equalizing phase and amplitude of the observed signal, the matched filter only equalizes the phase, but emphasizes the amplitude behavior of the signal, i.e. at its maximum, which corresponds to the location of the target, the output of the matched filter is the sum of the components of the observed signal, each weighted according to its own magnitude. Neglecting, for the moment, the influence of the antenna pattern and assuming the reflectivity to be constant over the aspect angle, (2.27) reveals that the magnitude of the received signal decreases as the magnitude of x and therefore the aspect angle α increases. Figure 2.4 shows a plot of the magnitude of s_x, normalized to its maximum,

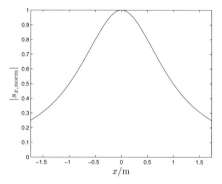

Figure 2.4: The normalized magnitude of s_x for aspect angles within $[-60°, 60°]$

along the synthetic aperture for aspect angles within $[-60°, 60°]$, received for a target in a range of $1\,\mathrm{m}$. The matched filter strengthens this effect, which means that reflectivity contributions from high aspect angles, that are expected to be weak, are even weakened by the matched filter. For targets with reflectivities that vary with the aspect angle, this means that the reflectivity contributions from different aspect angles are weighted non-uniformly. Common antenna patterns might even increase this effect, as the directivity often decreases as the aspect angle increases.

A variety of algorithms for the reconstruction of the reflectivity distribution from the observed signal exist. They basically differ in the assumptions and approximations they use in order to accelerate the processing. Among those algorithms are the *polar format* or *range Doppler* algorithm [29], the *range migration* or $\omega - k$ algorithm [30] and the *chirp scaling* algorithm [31], to name a few. Overviews and comparisons of them can be found in [32–34]. Another method, which is not based on approximations and does not require interpolations as some others do, is *range stacking* [35]. With range stacking, reflectivity maps are generated by matched-filtering the signals received for different frequencies, one at a time, and then adding the resulting single-tone reflectivity maps. The maps for each of the frequencies exhibit resolution in along-track direction. The summation of the processing results for different frequencies yields additional resolution in range-direction. Actually, even the single-tone processing results might exhibit resolution in range direction for sufficiently large observation angles, as will be shown later. That, however, is not a necessary prerequisite for image formation via range stacking.

The property exploited by all algorithms to obtain cross-range resolution is the (spatial) chirp-like nature of the signal recorded along the synthetic aperture. Correlating the recorded signal with a reference signal results in a compression of the spatially extended received signal into an output signal that is localized around the position of the target. As an example, consider the signal that is observable along the x-axis in response to a target with constant reflectivity at scene coordinates $(0, 1\,\mathrm{m}, 0)$, illuminated with an electromagnetic wave of the wavelength $\lambda = 12.5\,\mathrm{mm}$ according to a frequency of $24\,\mathrm{GHz}$. The antenna is assumed to have constant directivity within $\pm\alpha_0 = \pm 20°$ and zero directivity outside. The observed signal, whose real part is plotted in fig. 2.5(a),

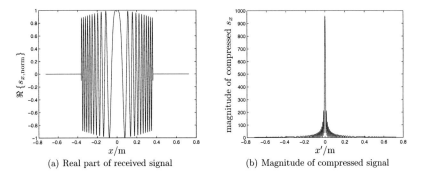

(a) Real part of received signal (b) Magnitude of compressed signal

Figure 2.5: Compressing the received signal into a cross-range image line

is non-zero at those positions, where the target is within the antenna beam, i.e. for $-0.36\,\mathrm{m} \le x_\mathrm{a} \le 0.36\,\mathrm{m}$. Compressing the received signal with a reference chirp yields a much narrower signal with a full width at half maximum (FWHM) of approximately $11\,\mathrm{mm}$, centered around the location of the target.

The following discussion of image properties will be closely related to the way an image can be generated using the range stacking algorithm. However, in contrast to the range stacking procedure, the single-tone images will not be generated by *matched* filters. Instead, the amplitude weighting is left as a degree of freedom, which can be used to control image properties like side-lobe levels, reflectivity weighting and the signal-to-noise ratio. Formulating the processing in a range-stacking-like manner is very illustrative and naturally linked to the type of data that is delivered by a stepped-frequency continuous-wave radar system as described in section 3. The obtained results are *not* limited to images generated via range stacking. They apply for all other algorithms as far as the influence of the involved approximations is properly accounted for.

As a basis for further discussion, *that* processing will be discussed, that is necessary to obtain an image, in which the magnitude of the pixel corresponding to the location of a (point-like) target is an estimate of the mean (amplitude) reflectivity of the target, where the reflectivities for different aspect angles are weighted *uniformly*. The mean value of the reflectivity within an aspect angle region $[-\alpha_0, \alpha_0]$, $\overline{\rho}_{\alpha_0}$, is

$$\overline{\rho}_{\alpha_0} = \frac{1}{2\alpha_0} \int\limits_{-\alpha_0}^{\alpha_0} \rho(\alpha) \mathrm{d}\alpha. \tag{2.35}$$

Note that for complex-valued reflectivity $\rho(\alpha)$, the mean value $\overline{\rho}_{\alpha_0}$ is to be understood as a complex number with real and imaginary parts equal to the mean of real and imaginary parts of $\rho(\alpha)$, respectively. The reflectivity $\rho(\alpha)$ for a given aspect angle can be found from the observed signal (2.27) by undoing whatever system and propagation have done, given that the range to the target is known. Therefore, according to the signal model established in section 2.3, $\overline{\rho}_{\alpha_0}$ can be determined according to

$$\overline{\rho}_{\alpha_0} = \frac{1}{2\alpha_0 a} \int\limits_{-\alpha_0}^{\alpha_0} s_\alpha(\alpha) \cdot \frac{r_\alpha^2(\alpha)}{D(\alpha)} \cdot \exp\left\{\mathrm{j}2kr_\alpha(\alpha)\right\} \mathrm{d}\alpha. \tag{2.36}$$

However, with a linear path of the radar system, data are recorded with constant spacing Δx in the x-domain rather than with constant spacing in the α-domain. Since a target at the scene coordinates $(0, r_0 \cos\beta, r_0 \sin\beta)$ is observed with an aspect angle α according to

$$\alpha = \arctan\left(\frac{-x}{r_0}\right) \tag{2.37}$$

from the radar system at $(x, 0, 0)$, the differential elements $\mathrm{d}\alpha$ and $\mathrm{d}x$ are related by

$$\mathrm{d}\alpha = \frac{-r_0}{r_x^2(x)} \mathrm{d}x. \tag{2.38}$$

Therefore, the unweighted mean reflectivity can be obtained from the observed signal according to

$$\overline{\rho}_{\alpha_0} = \frac{r_0}{2\alpha_0 a} \int\limits_{-x_0}^{x_0} s_x(x) \cdot \frac{1}{D(\alpha(x))} \cdot \exp\left\{\mathrm{j}2kr_x(x)\right\} \mathrm{d}x, \tag{2.39}$$

where x_0 is given by

$$x_0 = -r_0 \tan\alpha_0. \tag{2.40}$$

Comparing (2.36) and (2.39) reveals that the formulation in the x-domain does not contain an explicit compensation of the spreading loss according to r^2. That means, that when the synthetic aperture is sampled equidistantly—rather than equiangularly—the contributions from different aspect angles are weighted equally only when the spreading loss is *not* explicitly compensated for. This might seem contradictory, but is due to the fact that the number of samples per unit angle increases by exactly the same amount as the reflectivity contribution diminishes due to the higher spreading loss for increasing aspect angle. The varying spreading loss due to the varying range as observed along the synthetic aperture is inherently compensated for when the system's path is linear and samples are taken equidistantly. Therefore, (2.39) is the processing resulting in an *unweighted* mean of the target's reflectivity within the observed region of aspect angles for a *single* target supposed to be located at $(0, r_0 \cos\beta, r_0 \sin\beta)$.

In the preceding discussion, it was assumed that the position of the target was known. This, however, is commonly not the case, and it is the very purpose of SAR techniques to determine the positions of targets. For a target at scene coordinates $(x_t, r_0 \cos\beta, r_0 \sin\beta)$, the aspect angle α as seen from the antenna at $(x, 0, 0)$ is given by

$$\alpha = \arctan\left(\frac{x_t - x}{r_0}\right), \tag{2.41}$$

and the differential elements are related by

$$d\alpha = \frac{-r_0}{r_x^2(x_t - x)}dx. \tag{2.42}$$

The unweighted mean reflectivity of the target at the position x' at range r' within its region of aspect angles $[-\alpha_0, \alpha_0]$ can thus be obtained by modifying (2.39) as

$$\bar{\rho}_{\alpha_0}(x') = \frac{1}{2\alpha_0 a} \int_{x'-x_0}^{x'+x_0} s_x(x) \cdot \frac{r_x^2(x'-x)}{D(\alpha(x'-x))} \cdot \exp\left\{j2kr_x(x'-x)\right\} \cdot \frac{r'}{r_x^2(x'-x)}dx$$

$$= \frac{r'}{2\alpha_0 a} \int_{x'-x_0}^{x'+x_0} s_x(x) \cdot \frac{1}{D(\alpha(x'-x))} \cdot \exp\left\{j2kr_x(x'-x)\right\} dx. \tag{2.43}$$

Performing the processing according to (2.43) for a number of different x' and r' yields a 2-dimensional map with peaks at those pixels that correspond to the locations of (point) targets present in the scene, as indicated in fig. 2.5(b) and discussed in section 2.5.

In formulating (2.43), the uniformity of the weighting of reflectivity contributions from different aspect angles was emphasized. As this processing is not only suitable for estimating the mean reflectivity of an object at a known location, but also for finding the locations of objects present in the scene, it might be desirable to have a certain influence on image properties like side-lobe levels. It will be shown that controlling the side-lobe levels has immediate influence on the weighting of the reflectivity contributions for different aspect angles. This issue can be discussed by incorporating a *windowing* or *weighting* function w into (2.43), which then reads as

$$\iota_w(x') = \frac{r'}{2\alpha_0 a} \int\limits_{-\infty}^{\infty} s_x(x) \cdot \frac{w_x(x - x')}{D(\alpha(x' - x))} \cdot \exp\left\{ \mathrm{j}2kr_x(x' - x) \right\} \mathrm{d}x. \qquad (2.44)$$

$w_x(x)$ is assumed to be centered about $x = 0$ and to be non-zero only for $|x| \le |x_0|$, effectively limiting the range of integration to $[x' - x_0, x' + x_0]$ and allowing the integration limits to be extended to $\pm\infty$ without any change. $\overline{\rho}_{\alpha_0}$ has been replaced by ι_w, since a non-rectangular window changes the meaning of the processing result from an unweighted mean value, denoted by $\overline{\rho}$, to a general *image*, denoted by ι_w, where the subscript w reminds of the fact that a potentially non-rectangular window has been applied.

2.5 Image characteristics

In this section, resolution and side-lobe properties of SAR images generated according to (2.44) will be discussed. Resolution in cross-range and range direction will be evaluated. Cross-range resolution—and potentially even range resolution—depends on the limits of integration and the antenna's beam-width. However, not only the resolution depends on the beam-width, but also the content of the image may do so. Targets that are present in the scene might not even appear on the image in case they are oriented in such a way that they e.g. reflect the incident energy to directions from which they cannot be observed by the radar system along the synthetic aperture. The beam-width determines, which targets in a given scene can be observed by the radar system. The likelihood for a target not to be observable decreases with increasing beam-width. However, increasing the beam-width inevitably increases the influence of thermal noise in the SAR image as will be discussed in section 2.9. In order to determine the optimum beam-width for a given imaging task, a figure of merit describing the quality of the estimation will be introduced in section 2.8 which incorporates both effects mentioned. First, however, resolution and side-lobe issues will be discussed.

The full widths at half maximum (FWHM) of the main-lobe of the processing result will be considered as the system's resolutions in range and cross-range direction, δx and δr, respectively. Two targets with equal reflectivity, located at equal ranges, but spaced by a distance greater than δx in along-track direction, will result in two distinguishable lobes. It is important to note that for a coherent system the FWHM is to be measured for the amplitude of the processing result, not its power. The signal observed along the synthetic aperture is a superposition of the amplitudes—rather than the powers—of the signals echoed from the targets. With the coherent processing, the overall result is a superposition of the results that would be obtained for each of the targets alone. Therefore, the FWHM of the (amplitude) processing result corresponds to the $-6\,\mathrm{dB}$-width of the squared magnitude of the processing result, which is commonly used for display purposes.

2.5.1 Cross-range resolution and side-lobe levels

In case of matched filtering the observed signal, the processing result $\iota(x')$ equals the autocorrelation of the reference signal s,

$$\iota(x') = \int\limits_{-\infty}^{\infty} s(x)s^*(x-x')\mathrm{d}x. \tag{2.45}$$

Since the autocorrelation of s is related to the Fourier transform S of s via

$$\mathscr{F}\left\{ \int\limits_{-\infty}^{\infty} s(x)s^*(x-x')\mathrm{d}x \right\} = |S(k_x)|^2 \tag{2.46}$$

according to the Wiener-Kinchine theorem [36], the FWHM of the processing result can be determined when the magnitude of the spectrum of s is known. Processing data according to the weighted compensation approach (2.44), however, is *not* matched filtering, and therefore, the spectrum of the reference signal alone does not determine the processing result. That means that the knowledge about the imaging properties obtained with range-stacking as described in [35], which includes a matched filtering operation, cannot directly be applied. However, with reasonable assumptions and approximations, the processing can be written as an autocorrelation of an auxiliary signal derived from the reference signal and the compensation signal as will be done in the following, and thus resolution and side-lobe issues can be discussed in terms of the spectrum of that auxiliary signal. This is true at least for the region around the maximum of $\iota(x')$, which carries information on the resolution and on the levels of the first side-lobes. The discussion is closely related to that in [37].

For the determination of the resolution, the scene is assumed to consist of a single point-like target with constant reflectivity. Disregarding the factor $r'/(2\alpha_0)$, assuming a real-valued antenna pattern $D = D_A^2$, and substituting s_x according to (2.27) into (2.44), the processing result for a unit point-target is

$$\iota_w(x') =$$

$$\int\limits_{-\infty}^{\infty} \frac{D(\alpha(x))}{D(\alpha(x'-x))} \cdot \frac{1}{r_x^2(x)} \exp\left\{-\mathrm{j}2kr_x(x)\right\} \cdot w_x(x-x') \exp\left\{\mathrm{j}2kr_x(x'-x)\right\}\mathrm{d}x, \quad (2.47)$$

whose shape determines resolution and side-lobe levels, and which therefore will be evaluated in the following.

For an antenna pattern that varies slowly around $\alpha = 0$, the ratio of the antenna pattern and its shifted version in (2.47) is approximately equal to 1. Numerical investigations at the end of this section will show that this is a valid assumption for a variety of patterns. Then, the processing result is approximately

$$\iota_w(x') = \int\limits_{-\infty}^{\infty} \frac{1}{r_x^2(x)} \exp\left\{-\mathrm{j}2kr_x(x)\right\} \cdot w_x(x-x') \exp\left\{\mathrm{j}2kr_x(x'-x)\right\}\mathrm{d}x, \tag{2.48}$$

which is the cross-correlation of two chirp-like signals exhibiting equal *non-linear* frequency modulation, but different amplitude modulations. The first chirp is inherently modulated by

$$w_{1,x}(x) = \frac{1}{r_x^2(x)} = \frac{1}{r_0^2 + x^2}. \tag{2.49}$$

The second one is modulated by the weighting function chosen by the system designer to influence the imaging result,

$$w_{2,x}(x) = w_x(x). \tag{2.50}$$

Since the resolution, δx, is supposed to be very small as compared to x_0, and since common windowing functions vary only very few over this small portion of their extent, it seems reasonable to assume that

$$w_{1,x}(x - x') = w_{1,x}(x), \quad w_{2,x}(x) = w_{2,x}(x - x') \tag{2.51}$$

approximately holds true for small values of x'. Using this assumption, the processing result can be rewritten as

$$\iota_w(x') = \int_{-\infty}^{\infty} w_{12,x}(x) \exp\left\{-\mathrm{j}2kr_x(x)\right\} \cdot w_{12,x}(x - x') \exp\left\{\mathrm{j}2kr_x(x - x')\right\} \mathrm{d}x, \tag{2.52}$$

being the autocorrelation of a non-linear chirp with an amplitude modulation according to

$$w_{12,x}(x) = \sqrt{w_{1,x}(x) \cdot w_{2,x}(x)}. \tag{2.53}$$

Thus, according to (2.46), $\iota_w(x')$ can be approximated as the inverse Fourier transform of the power spectral density of the amplitude and phase modulated auxiliary signal

$$c(x) = w_{12,x}(x) \exp\left\{-\mathrm{j}2kr_x(x)\right\} \tag{2.54}$$

in the vicinity of $x' = 0$. The spectrum of such a signal can approximately be and commonly is determined by the method of stationary phase. However, since only the magnitude of the spectrum rather than its phase is of interest to determine $\iota_w(x')$, the following very illustrative derivation is sufficient.

For sufficiently large time-bandwidth product—or space-spatial bandwidth product as might be a more appropriate term, since $c(x)$ is a function of the spatial coordinate rather than time—the spectral contents of $c(x)$ can be assumed to be determined exclusively by the relative contributions of the instantaneous (spatial) wave-numbers. The phase of the exponential term in (2.54),

$$\Phi(x) = -2kr_x(x) = -2k\sqrt{r_0^2 + x^2}, \tag{2.55}$$

changes with x. The instantaneous (spatial) wave-number, k_x, at the position x is the derivative of the phase with respect to x,

$$k_x(x) = \frac{\mathrm{d}\Phi(x)}{\mathrm{d}x} = \frac{-2kx}{\sqrt{r_0^2 + x^2}}. \tag{2.56}$$

Thus, the location x corresponding to a given wave-number k_x is

$$x(k_x) = \pm \frac{k_x r_0}{\sqrt{4k^2 - k_x^2}}. \tag{2.57}$$

The magnitude of the variation of x with k_x,

$$\left| \frac{\mathrm{d}x(k_x)}{\mathrm{d}k_x} \right| = \frac{4r_0 k^2}{(4k^2 - k_x^2)^{\frac{3}{2}}}, \tag{2.58}$$

can be used to determine the power contained in the unweighted exponential term around a certain wave-number. Figuratively spoken, the higher the variation of x for a certain variation of k_x, the wider the region in which a certain wave-number is present and therefore the higher the power spectral density for this wave-number. However, not only the wave-number varies with x. Also, the amplitude varies like w_{12}, and therefore, the power is influenced by w_{12}^2. Thus, the (relative) power spectral density can be approximated (in terms of k_x) by

$$\left| \tilde{C}(k_x) \right|^2 = \frac{4r_0 k^2}{(4k^2 - k_x^2)^{\frac{3}{2}}} \cdot w_{12,k_x}^2(k_x) \tag{2.59}$$

within its support band and zero outside, where

$$w_{12,k_x}(k_x) = w_{12,x}(x(k_x)) \tag{2.60}$$

is the scale transformed version of a weighting function originally given in the x-domain. The support band of the signal is within $\pm|k_{x,0}|$, where

$$k_{x,0} = k_x(\alpha_0) = -2k \sin \alpha_0. \tag{2.61}$$

Therefore, the width of the spectrum is

$$\Delta_{k_x} = 4k \sin \alpha_0. \tag{2.62}$$

Note that (2.59) is relative in that it is not normalized to the signal power. However, resolution and relative side-lobe levels are invariant with respect to a constant factor. Finally, $\iota_w(x')$ can be approximated by

$$\begin{aligned} \iota_w(x') &= \mathscr{F}^{-1} \left\{ \frac{4r_0 k^2}{(4k^2 - k_x^2)^{\frac{3}{2}}} \cdot w_{12,k_x}^2(k_x) \right\} \\ &= \mathscr{F}^{-1} \left\{ w_{i,k_x}(k_x) \cdot w_{2,k_x}(k_x) \right\}, \end{aligned} \tag{2.63}$$

where

$$w_{i,k_x}(k_x) = \frac{1}{r_0 \sqrt{4k^2 - k_x^2}} \tag{2.64}$$

will be referred to as the *intrinsic* window caused by the non-linear frequency modulation with respect to x and the amplitude modulation inherent in the imaging geometry.

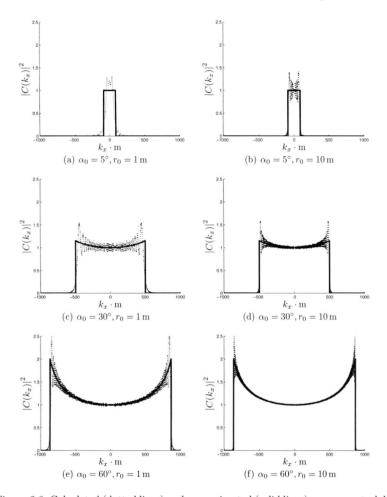

Figure 2.6: Calculated (dotted lines) and approximated (solid lines) power spectral densities of $c(x)$ for various combinations of α_0 and r_0.

w_2 can be chosen arbitrarily to influence the properties of the processing result. Figure 2.6 shows plots of the power spectral densities of auxiliary signals $c(x)$ according to different beam-widths and target ranges—calculated as the discrete Fourier transforms and approximated according to (2.59)—for w_2 being rectangular, i.e. for the case of uniform reflectivity weighting. As beam-width and target range are increased, the time-bandwidth product is increased and therefore the quality of the approximation improves.

Additionally, the shape of the approximation changes as the beam-width changes. For small integration angles, w_i is almost constant, i.e. the power spectral density is almost rectangular. Therefore, the width of the main-lobe and the levels of the side-lobes are determined approximately by the inverse Fourier transform of w_{2,k_x} only. Choosing w_2 to be a rectangular window (uniform reflectivity weighting), the magnitude of $\iota_w(x')$ exhibits a sinc-shape with the first side-lobes 13 dB below the maximum. The FWHM of this sinc-shape and therefore the cross-range resolution δ_x is

$$\delta_x = 0.3 \cdot \frac{\lambda}{\sin \alpha_0} \tag{2.65}$$

according to the width Δ_{k_x} of the window in k_x-domain, where λ is the wavelength of the radiated signal.

In contrast to a radar system with a real aperture, the cross-range resolution of a SAR system does not depend on the range of the target. However, the length of the synthetic aperture needed to cover a certain region of aspect angles increases with increasing range. Note that the resolution predicted by (2.65) is obtainable only for a target whose reflectivity is constant over the aspect angle. A target that is only partially observable along the synthetic aperture will produce inferior resolution due to the reduced spatial bandwidth of the received signal.

For other than rectangular windows, δ_x is increased according to the ratio of the 6 dB-bandwidths of the respective window and that of the rectangle. As mentioned before, the 6 dB-bandwidth is relevant since the amplitudes rather than the powers of the signals received from reflexions of several targets are superimposed in the receiver. It is also stated in [38] that the 6 dB-bandwidth rather than the 3 dB-bandwidth determines the resolution in coherent systems.

For the narrow antenna beams commonly used in long-range SAR systems, the approximation $\sin \alpha_0 \approx \tan \alpha_0 \approx \alpha_0$ is valid. In this case, the antenna beam-width $2\alpha_0$ is commonly approximated as

$$2\alpha_0 = a_{\text{tf}} \frac{\lambda}{l}, \tag{2.66}$$

with l the length of the antenna's real aperture in along-track direction and a_{tf} the aperture illumination taper factor [39, p. 646] of the antenna. In this case, (2.65) reduces to

$$\delta_x = \frac{0.3}{a_{\text{tf}}} \cdot l, \tag{2.67}$$

which is qualitatively consistent with the expression commonly used for the resolution as a function of the length of the real antenna aperture, given e.g. in [39]. The factor 0.3 results from defining the resolution as the FWHM and might to be replaced by a different factor for a different definition of the resolution. While (2.67) is sufficient to determine the resolution of narrow-beam systems in terms of the spatial extent of their antennas, the resolution for wide beam systems is given by (2.65) in terms of the integration angle, but only in case the intrinsic window is compensated for, as will be discussed next.

As the integration angle is increased, w_i is no longer approximately constant, as can be seen in fig. 2.6. Numerical simulations show that the level of the first side-lobes increases with increasing integration angle and the resolution improves slightly. Figure

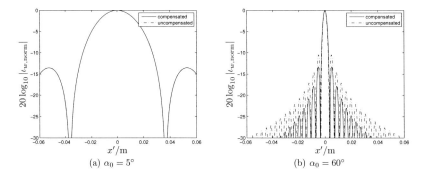

(a) $\alpha_0 = 5°$ (b) $\alpha_0 = 60°$

Figure 2.7: 10-m-range-lines for compensated and uncompensated intrinsic windows

2.7 shows the 10-m-range-lines generated for a target at a range of 10 m with antennas with $\alpha_0 = 5°$ and $\alpha_0 = 60°$, respectively, processed with and without compensation of the intrinsic window. For $\alpha_0 = 5°$, there is no visual difference between the results obtained with and without compensation of the intrinsic window. Both exhibit equal widths of the main-lobes and therefore equal resolutions. The level of the first side-lobes is $-13\,\mathrm{dB}$ with respect to the maximum, as expected for rectangular weighting. For $\alpha_0 = 60°$, the first side-lobes are $13\,\mathrm{dB}$ below the maximum in case the intrinsic window is compensated for, but only $11\,\mathrm{dB}$ below the maximum without compensation. Additionally, the resolution obtained without compensation is slightly different from the one obtained with compensation.

The simulation result in fig. 2.7(a) for $\alpha_0 = 5°$ and $r_0 = 1\,\mathrm{m}$ shows that the approximations that led to the formulation of (2.63) are well suited to predict the behavior of the processing result around the location of the target, even if the approximation of the power spectral density according to (2.59) is far from being perfect for such small integration angles, target ranges and the resulting comparatively low time-bandwidth products.

In order to obtain equal weights for reflectivity contributions from all observed angles, the intrinsic window *must not* be compensated for, since that would introduce a weighting according to the inverse of (2.68). Side-lobe levels, that increase as α_0 increases, are a characteristic property of estimating the unweighted mean of the reflectivity. However, when the uniformity of the weighting is not an issue, it can be rendered in favor of influence on the side-lobe levels. Amongst others, Harris [38] has given a table containing information on the properties of the Fourier transforms of a large number of windows. Those tabulated properties are valid for the imaging system as far as they are related to the vicinity of the main-lobe—so that assumption (2.51) is valid—and the time-bandwidth product of (2.54) is high enough—so that approximation (2.59) is valid. Such can be resolution and the levels of the first side-lobes. Furthermore, α_0 has to be sufficiently large so that the influence of the variation of D around $\alpha = 0$ can be neglected. In applying windows for high integration angles, two things have to be taken care of.

First, for the processing result to be equal to the inverse Fourier transform of the desired window, the power spectral density of the auxiliary function c needs to have the shape of that desired window, which is accomplished by multiplying the desired window with $1/w_i$. This compensation inevitably weights the reflectivity contributions like $\cos \alpha$, since with (2.64), (2.56) and $\sin^2 \alpha = x^2/(r_0^2 + x^2)$, the intrinsic window reads as

$$w_{i,\alpha}(\alpha) = \frac{1}{2kr_0 \cos \alpha} \tag{2.68}$$

in the α-domain.

Second, since the image is equal to the inverse Fourier transform of the power spectral density with respect to k_x, the window has to be established with respect to k_x. For small integration angles, i.e. $x_{\max} \ll r_0$, relation (2.56) between k_x and x is approximately linear, and therefore, weighting in the x-domain is approximately equal to weighting in the k_x-domain. However, as the integration angle becomes larger, the relation becomes non-linear, and therefore, the window has to be established over k_x and then non-linearly scale transformed in order to be applied to the x-domain data. A Hann window will be used to illustrate this matter.

$$w_{\text{Hann},k_x}(k_x) = \cos^2 \left(\frac{k_x}{k_{x,0}} \cdot \frac{\pi}{2} \right) \tag{2.69}$$

for $|k_x| \leq k_{x,0}$ and zero otherwise is a Hann window in the k_x-domain. The corresponding weighting function to be applied in the x-domain is

$$w_{\text{Hann},x}(x) = w_{\text{Hann},k_x}(k_x(x)), \tag{2.70}$$

where the scale transform is given by (2.56) and results in

$$w_{\text{Hann},x}(x) = \cos^2 \left(\frac{x}{x_0} \cdot \frac{\sqrt{r_0^2 + x_0^2}}{\sqrt{r_0^2 + x^2}} \cdot \frac{\pi}{2} \right). \tag{2.71}$$

Fig. 2.8 shows a k_x Hann window, scale transformed to the indicated domains, for $\alpha_0 = 60°$. It is clearly visible that the shape of what acts as a Hann window in the k_x-domain is different in x- and α-domain. ξ is a placeholder for k_x, α and x in the respective domains, and $k_{x,0}$, x_0 and α_0 are related by

$$k_{x,0} = -2k \sin \alpha_0$$

and

$$x_0 = -r_0 \tan \alpha_0.$$

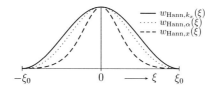

$$w_{\mathrm{Hann},k_x}(\xi)$$
$$w_{\mathrm{Hann},\alpha}(\xi)$$
$$w_{\mathrm{Hann},x}(\xi)$$

Figure 2.8: k_x Hann window plotted in k_x, α and x-domain

Considering the compensation of the intrinsic window, the actual window to be applied is

$$w_{2,x}(x) = \frac{w_{\mathrm{Hann},x}(x)}{w_{\mathrm{i},x}(x)} = \frac{2kr_0^2}{\sqrt{r_0^2 + x^2}} \cdot w_{\mathrm{Hann},x}(x). \qquad (2.72)$$

Note that the behavior far apart from the main-lobe may differ significantly from the behavior of the Fourier transform of a Hann window, since the approximations made in order to determine resolution and side-lobe levels are valid only close around the main-lobe. However, in the vicinity of the main-lobe, the imaging result is closely related to the Fourier transform of the applied window.

From fig. 2.8, it is obvious that windowing in the x-domain not only influences the spectrum in the k_x-domain and therefore resolution and side-lobe levels, but also the weighting of reflectivity contributions from different aspect angles. For uniform reflectivity weighting, the window to be applied is rectangular in either domain. However, the resulting spectrum is approximately rectangular only for narrow antenna beams, and its shape changes as the beam-width increases. As a consequence, the side-lobes raise. If not the uniform weighting of the reflectivities but a constant side-lobe level is desired, the spectrum has to be equalized by compensating for the intrinsic window, resulting in a $\cos\alpha$-weighting of the reflectivity contributions. The actual side-lobe levels can then be influenced by applying the proper weighting function in addition to compensation of the intrinsic window.

This section on the cross-range resolution and side-lobe properties will be concluded with simulation results giving an impression under which circumstances the assumption made to formulate (2.48)—i.e. that the antenna pattern is almost perfectly compensated even by a slightly shifted pattern compensation function—is feasible to predict the behavior of the processing result close to the main-lobe. In section 2.9.4, those antenna patterns will be derived that minimize the influence of thermal noise on the processing result for given weighting functions. The optimum patterns for an uncompensated rectangular window and compensated Hann and Blackman windows are plotted in fig. 2.31 for $\alpha_0 = 60°$. A major difference between the pattern optimized for a rectangular window on the one hand and for a Hann and a Blackman window on the other hand is their behavior as α_0 is varied. For the rectangular window, there is no change in the shape of the optimum pattern. Decreasing α_0 will just truncate the given pattern at the new value of α_0. For the Hann and the Blackman window, however, the shape will change. The pattern will be compressed to fit into $[-\alpha_0, \alpha_0]$ for the new value of α_0,

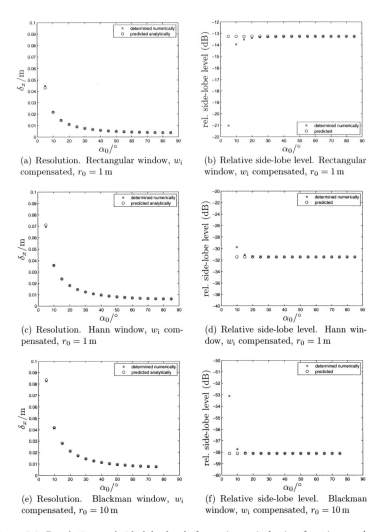

(a) Resolution. Rectangular window, w_i compensated, $r_0 = 1\,\mathrm{m}$

(b) Relative side-lobe level. Rectangular window, w_i compensated, $r_0 = 1\,\mathrm{m}$

(c) Resolution. Hann window, w_i compensated, $r_0 = 1\,\mathrm{m}$

(d) Relative side-lobe level. Hann window, w_i compensated, $r_0 = 1\,\mathrm{m}$

(e) Resolution. Blackman window, w_i compensated, $r_0 = 10\,\mathrm{m}$

(f) Relative side-lobe level. Blackman window, w_i compensated, $r_0 = 10\,\mathrm{m}$

Figure 2.9: Resolution and side-lobe levels for various windowing functions and corresponding optimized antenna patterns

which means that the curvature around $\alpha = 0$ will increase, and therefore the quality of the compensation of the antenna pattern by a slightly shifted version will decrease.

Figure 2.9 shows comparisons of resolutions and side-lobe levels, as obtained by simulations and predicted analytically in this section. Comparisons are available for rectangular, Hann and Blackman windows, where the intrinsic window has been compensated with all three of them. For the rectangular window with its almost flat optimum pattern around $\alpha = 0$ (irrespective of α_0), resolution and side-lobe levels are predicted almost perfectly even for antenna beams as narrow as $\pm 5°$ and a target range of $1\,\mathrm{m}$, as depicted in fig. 2.9(a) and fig. 2.9(b). For a Hann window, with its optimum pattern being increasingly curved around $\alpha = 0$ for small α_0, the resolution is predicted fairly well starting from relatively small α_0 (fig. 2.9(c)). The prediction of the relative level of the first side-lobe is correct to within $1\,\mathrm{dB}$ starting around $\alpha_0 = 15°$ (fig. 2.9(d)). Qualitatively the same can be read off the plots for Blackman windows in fig. 2.9(e) and fig. 2.9(f), but only because the target range has been increased to $r_0 = 10\,\mathrm{m}$. For a range of $r_0 = 1\,\mathrm{m}$, the approximation is feasible for larger α_0 only. Figure 2.10

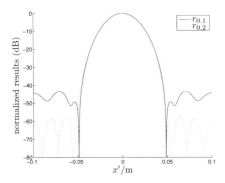

Figure 2.10: Processing results obtained for Blackman window with optimized antenna pattern for $\alpha_0 = 10°$, $r_{0,1} = 1\,\mathrm{m}$, $r_{0,2} = 10\,\mathrm{m}$

shows a comparison of the processing results obtained with a Blackman window and its optimized pattern for $\alpha_0 = 10°$ for target ranges $r_{0,1} = 1\,\mathrm{m}$ and $r_{0,2} = 10\,\mathrm{m}$. The quality of the approximation improves as the target range increases. This has two reasons. First, the (spatial) time-bandwidth product of the received signal increases. Second, for a close target, the antenna pattern, being a function of α, varies faster as the antenna moves along the synthetic aperture than it does for a distant target. Therefore, the pattern as effective for a distant target does not vary as much around $\alpha = 0$ within a given cross-range region as it does for a target close to the synthetic aperture. Consequently, the approximation becomes valid within a region of increasing extent around $x' = 0$ as the target range increases.

It is well-known that windowing functions, applied in order to influence e.g. side-lobe levels, have impact on the signal-to-noise ratio of the processing result at the position of the target. This impact, and antenna patterns that minimize the thermal noise contributions for given windowing functions, will be discussed in section 2.9.4. First, however, resolution issues with respect to the range-direction will be addressed.

2.5.2 Range resolution for zero-bandwidth signal

In the discussion of cross-range resolution, only the range line corresponding to the range of the target has been considered. Investigating the behavior of the processing result at the cross-range coordinate of the target, but for variable range, reveals that even for a system that emits a purely sinusoidal signal, resolution in range direction can be obtained. This effect is rarely mentioned in SAR literature, presumably for it requires comparatively large aspect angle variations and targets that are observable over such large regions of aspect angles. The coverage of a large aspect angle region demands for narrow sample spacing along the synthetic aperture, as will be discussed in section 2.6. Additionally, depending on the extent of the generated image, the imaging result might exhibit only a *local* rather than a global maximum at the location of the target, since the magnitude of the single-tone processing result increases with increasing range, as will be shown in this section. For some applications, those properties might not pose limitations, allowing for low-budget system implementations and complying with potentially rigorous bandwidth regulations. In [40, p. 104], range resolution for a single-tone signal is derived based on a consideration of phase errors. It is stated that this derivation is valid only for aperture angles less than 1 rad. However, a resolution estimation based on the k_r bandwidth of the received signal, which is a function of α_0, yields the same expression for the resolution, and a comparison with the resolution obtained by numerical simulations shows that the expression is a fair measure for the resolution even as α_0 approaches 90°.

k_x as defined in (2.56) and written in terms of α is the length of the projection of the wave-vector of the transmitted and received signal with angle of incidence α_0 onto the x-axis, where the length of the effective wave-vector is $2k$ due to the two-way propagation to the target and back. The length of the wave-vector's projection onto a plane perpendicular to the synthetic aperture, being the radial component of the wave-vector, is

$$k_r = 2k \cos \alpha. \tag{2.73}$$

As α varies between $-\alpha_0$ and α_0, k_r varies between $k_{r,\mathrm{min}}$ and $k_{r,\mathrm{max}}$ according to

$$k_{r,\mathrm{min}} = 2k \cos \alpha_0 \tag{2.74}$$

and

$$k_{r,\mathrm{max}} = 2k \cos 0 = 2k. \tag{2.75}$$

Therefore, the k_r-bandwidth covered as α varies within $[-\alpha_0, \alpha_0]$ is

$$\Delta_{k_r} = k_{r,\mathrm{max}} - k_{r,\mathrm{min}} = 2k(1 - \cos \alpha_0). \tag{2.76}$$

The corresponding resolution, defined as the FWHM as with the cross-range direction, is

$$\delta_r = \frac{0.6\lambda}{1 - \cos \alpha_0}, \tag{2.77}$$

which is, up to a constant, equal to the expression given in [40, p. 104]. The same behavior, up to a constant, is known for the axial resolution of confocal microscopes, as documented e.g. in [41, p. 209]. In [42], a resolution area for *wide-band* wide-beam systems is derived. As part of this derivation, k_r is also determined, but the extent of its spectrum is not used to determine the range resolution for the case of a wide-beam system with *zero temporal bandwidth*. Rather, it is stated that for a given beam-width there is a minimum temporal signal bandwidth that needs to be covered in order to obtain a reasonable integrated mainlobe-to-sidelobe energy ratio (ISLR). That bandwidth necessary is stated in section 2.5.3.

Before comparing the zero-bandwidth range resolution predicted by (2.77) with simulation results, the behavior of the processing result apart from the location of the target will be investigated.

The method of stationary phase (MSP), which is discussed e.g. in [43] and summarized e.g. in [44], can be used to approximate the range-behavior of the processing result apart from the location of the target in case of signals that oscillate fast enough over a sufficient range of frequencies—a prerequisite that may be fulfilled for sufficiently large beam-width apart from the location of the target. According to the MSP, the *magnitude* of

$$\int g(x) \exp\{j\Phi(x)\} \mathrm{d}x \tag{2.78}$$

is approximated by

$$\sqrt{\frac{2\pi}{\Phi''(x^*)}} \cdot g(x^*), \tag{2.79}$$

where x^* is the *stationary point* defined by

$$\Phi'(x^*) = 0. \tag{2.80}$$

Applying the MSP, one assumes that the value of the integral is determined by the value of the function f and the second derivative of the phase function Φ at the stationary point only. For the cross-range location of the target with constant reflectivity ρ, the processing result at range $r_0 + \Delta r$ for a target at range r_0 reads as

$$\iota(r_0, \Delta r) = \frac{r_0 + \Delta r}{2\alpha_0} \int_{-r_0 \tan \alpha_0}^{r_0 \tan \alpha_0} \frac{\rho}{r_0^2 + x^2} \exp\left\{-j2k\sqrt{r_0^2 + x^2}\right\}$$
$$\cdot \frac{w(x, r_0 + \Delta r)}{w_{\mathrm{i}}(x, r_0 + \Delta r)} \exp\left\{j2k\sqrt{(r_0 + \Delta r)^2 + x^2}\right\} \mathrm{d}x. \tag{2.81}$$

The stationary point of the associated phase function

$$\Phi(x) = 2k\left(\sqrt{(r_0 + \Delta r)^2 + x^2} - \sqrt{r_0^2 + x^2}\right) \tag{2.82}$$

is found by equating

$$\Phi'(x) = 2kx \cdot \left[\frac{1}{\sqrt{(r_0 + \Delta r)^2 + x^2}} - \frac{1}{\sqrt{r_0^2 + x^2}}\right] \tag{2.83}$$

to zero to be

$$x^* = 0. \tag{2.84}$$

Therefore, the magnitude of the imaging result can be approximated according to (2.79) by

$$|\iota_{\mathrm{MSP}}(r_0, \Delta r)| = \frac{\rho}{2\alpha_0} \cdot \sqrt{\frac{\pi}{kr_0^3}} \cdot \frac{w(0, r_0 + \Delta r)}{w_\mathrm{i}(0, r_0 + \Delta r)} \cdot \frac{(r_0 + \Delta r)^{3/2}}{\sqrt{\Delta r}}, \tag{2.85}$$

which for $\Delta r \gg r_0$ and $r = r_0 + \Delta r \approx \Delta r$ reduces to

$$|\iota_{\mathrm{MSP}}(r_0, r)| = \frac{\rho}{2\alpha_0} \cdot \sqrt{\frac{\pi}{kr_0^3}} \cdot \frac{w(0, r_0 + \Delta r)}{w_\mathrm{i}(0, r_0 + \Delta r)} \cdot r. \tag{2.86}$$

For determining the intrinsic window w_i in section 2.5.1, the factor r' has been disregarded, since there only a single range line has been considered and disregarding this factor made the formulation of cross-range imaging in (2.47) easier while still being correct in the respective context. Since here the processing result is evaluated for different ranges, the range factor needs to be included. Re-including this factor, the value of the intrinsic window for $x = 0$ is not a function of range. The magnitudes of windows used to influence side-lobe properties are independent of range for $x = 0$, too. Therefore, the ratio of w and w_i is a constant in (2.85) and (2.86). The higher this ratio, the steeper the slope of the imaging result with respect to r. However, the ratio of w and w_i—to be precise: its integral—determines the height of the peak at the position of the target. Depending on the shape of the windows, the peak height will be multiplied by a factor on the order of 0.5 for common windowing functions. Assuming a rectangular window without compensation of the intrinsic window, the imaging result as approximated by the MSP and normalized to ρ reads as

$$|\iota_{\mathrm{MSP,norm}}(r_0, \Delta r)| = \frac{1}{2\alpha_0} \cdot \sqrt{\frac{\pi}{kr_0^3}} \cdot \frac{(r_0 + \Delta r)^{3/2}}{\sqrt{\Delta r}}, \tag{2.87}$$

and

$$|\iota_{\mathrm{MSP,norm}}(r_0, r)| = \frac{1}{2\alpha_0} \cdot \sqrt{\frac{\pi}{kr_0^3}} \cdot r, \tag{2.88}$$

respectively. Figure 2.11 shows 24-GHz single-tone processing results as obtained numerically (solid lines) and according to the MSP (dotted) approximation (2.87) for varying r_0 and α_0. For $\alpha_0 = 10°$ and a target at range $r_0 = 0.1\,\mathrm{m}$ (fig. 2.11(a)), there is no distinctive peak at the reconstructed range $r = 0.1\,\mathrm{m}$, i.e. single-tone processing does not deliver resolution for such small integration angles and short target ranges. As either the target range (fig. 2.11(b)) or the integration angle (fig. 2.11(c)) are increased, it makes sense to define a range-resolution. However, the target lobe is not necessarily the maximum in the reconstructed image, as the magnitude of the processing result increases with r, where the slope is inversely proportional to the integration angle, to the square root of the wave number, and to $r_0^{2/3}$. The given dependence of the slope on α_0 can be seen from comparing the plots for different integration angles but equal ranges r_0, whereas the dependence on r_0 becomes visible from comparing the plots

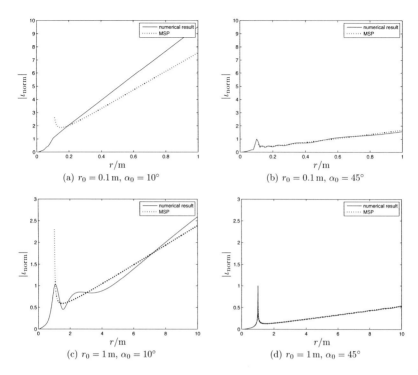

Figure 2.11: 24-GHz single-tone processing results as obtained numerically (solid lines) and according to the MSP (dotted) approximation (2.87) for varying r_0 and α_0

for equal α_0 but different r_0. Those plots reveal that a very weak but close target can mask a strong target that is in a far greater distance. Note that this problem cannot be resolved by just dividing the processing result by r, since then the peak of the remote target would also be divided by r.

Also obvious from the plots in fig. 2.11 is the improving resolution for increasing integration angles. Equation (2.77) predicts the obtainable single-tone range-resolution for given α_0. Figure 2.12 shows plots comparing the numerically determined resolution and the predicted one for a target at range $r_0 = 5\,\mathrm{m}$ for a variety of integration angles. Those plots, showing results obtained without and with compensation of the intrinsic window, reveal that (2.77) is a fair measure for the single-tone range-resolution even for large α_0.

However, not the beam-width of the antenna alone, but also the region of aspect angles within which a target is observable, determine the obtainable single-tone range-

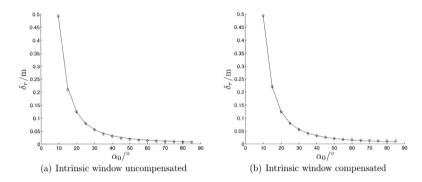

(a) Intrinsic window uncompensated   (b) Intrinsic window compensated

Figure 2.12: Range-resolution as predicted (solid line) by (2.77) and determined numerically (diamonds)

resolution. This issue becomes relevant when the objects to be imaged are directive, as e.g. metal plates can be. Such targets are observable only along a part of the synthetic aperture, resulting in a narrower support band of k_r and therefore inferior resolution as compared to a target that exhibits constant reflectivity over the observed region of aspect angles. Figure 2.13 shows a -6-dB-contour plot of a hypothetic scenery consisting of two targets, imaged at a single frequency point (24 GHz). The first one is a target with constant reflectivity, located at the scene coordinates $(0.4\,\mathrm{m}, 1\,\mathrm{m}, 0)$. The second one exhibits non-zero constant reflectivity for aspect angles within $\pm 10°$ and zero reflectivity outside and is located at $(-0.4\,\mathrm{m}, 1\,\mathrm{m}, 0)$. The contours have been determined by processing the signals expected for an antenna with a beam-width of $2\alpha_0 = 60°$ and compensation of the intrinsic window. The omni-directional target is imaged with range- and cross-range resolutions as expected for the given beam-width. The directive target, however, yields range and cross-range resolutions inferior to that of the omni-directional target due to its reduced observability along the synthetic aperture and the resulting lower k_x- and k_r-bandwidths. It can be seen that the extent of the directive target's contour is not maximum at $r' = 1\,\mathrm{m}$. The single-tone cross-range resolution as defined in (2.65) is the width of the contour at the range of the target rather than its maximum extent in cross-range direction when a single frequency is used.

The loss in range-resolution can be circumvented by covering a non-zero (temporal) signal bandwidth, either by transmitting a short pulse or equivalently by using harmonic signals of different frequencies one after the other, which is the mode discussed in the following sections. Non-zero bandwidths will also reduce the extent of the contour for ranges greater than that of the target, yielding a contour with maximum cross-range extent at the range of the target, giving (2.65) the meaning of an effective resolution in cross-range direction.

Summarizing the results of this section, it can be stated that the range-behavior of

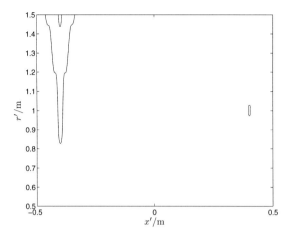

Figure 2.13: −6-dB-contour plot for partially observable (left) and fully observable (right) targets

single-tone imaging is fairly described by a lobe with an FWHM extent according to (2.77) around the location of the target and a slope according to (2.87) for reconstructed distances greater than that of the target, potentially exceeding the level of the target lobe. That slope, which depends on parameters of the imaging system and on the location of the targets, might be tolerable for some applications, allowing for cost-effective narrow-band implementations. Otherwise, this effect can be counteracted by increasing the bandwidth of the transmitted signal above zero.

2.5.3 Range resolution for non-zero-bandwidth signal

For comparatively narrow antenna beams, the range resolution of a SAR system is determined by the bandwidth of the transmitted signal only, and it is given by (s. e.g. [35])

$$\delta r = \frac{c}{2B}, \tag{2.89}$$

where c is the propagation velocity of the electromagnetic wave, and B the bandwidth of the transmitted signal. In what follows, it will be shown how merely adding the cross-range processing results obtained for different frequencies yields resolution in range direction. This procedure is known as *range stacking* and discussed in [35] for a continuum of frequencies. Here, it will be discussed for a discrete frequency spectrum, closely related to the type of data obtained with a stepped-frequency continuous-wave (SFCW) SAR system. It is shown in [45], that by performing an inverse discrete Fourier transform (IDFT) on the coherent data obtained for different frequencies within a certain bandwidth, a range profile of the scenery can be generated—very similar to the range profile that is obtained by using a pulsed or chirped radar system with the same bandwidth.

The processing to be investigated is a summation (and division by N) of the cross-range processing results obtained for N different frequencies f_n— starting at $f_{\min}$ and spaced by f_Δ—and the corresponding wave-numbers

$$k_n = \frac{2\pi f_n}{c} = \frac{2\pi(f_{\min} + nf_\Delta)}{c} \tag{2.90}$$

and reads as (according to (2.81) without considering windowing functions with respect to x)

$$\iota_N(r_0, \Delta r) = \frac{1}{N} \sum_{n=0}^{N-1} \frac{r_0 + \Delta r}{2\alpha_0} \cdot$$

$$\int_{-r_0 \tan \alpha_0}^{r_0 \tan \alpha_0} \frac{\rho}{r_0^2 + x^2} \exp\left\{-j2k\sqrt{r_0^2 + x^2}\right\} \cdot \exp\left\{j2k\sqrt{(r_0 + \Delta r)^2 + x^2}\right\} dx \tag{2.91}$$

for the cross-range coordinate of the target with constant reflectivity ρ. Assuming a narrow antenna beam, x remains approximately constant, and $\tan \alpha_0$ can be approximated by α_0. Then, the overall processing result can be approximated as

$$\iota_N(r_0 + \Delta r) = \rho \cdot \frac{r_0 + \Delta r}{r_0} \cdot \frac{1}{N} \sum_{n=0}^{N-1} \exp\left\{j2\frac{2\pi(f_{\min} + nf_\Delta)}{c}\Delta r\right\}. \tag{2.92}$$

Discretizing the range direction according to

$$r_0 + \Delta r = r_0 + d \cdot r_\Delta, \tag{2.93}$$

the discretized processing result reads as

$$\iota_N[d] = \rho \cdot \frac{r_0 + d \cdot r_\Delta}{r_0} \cdot \exp\left\{j\frac{4\pi}{c}f_{\min}dr_\Delta\right\} \cdot \frac{1}{N} \sum_{n=0}^{N-1} \exp\left\{j2\pi nd \cdot \frac{2f_\Delta r_\Delta}{c}\right\}. \tag{2.94}$$

Requiring that

$$N \cdot \frac{2f_\Delta r_\Delta}{c} = 1, \tag{2.95}$$

the above can be written as the IDFT of a unit function, multiplied by an amplitude and a phase term:

$$\iota_N[d] = \rho \cdot \frac{r_0 + d \cdot r_\Delta}{r_0} \cdot \exp\left\{j2\pi \cdot \frac{f_{\min}}{f_\Delta} \cdot \frac{d}{N}\right\} \cdot \frac{1}{N} \sum_{n=0}^{N-1} \exp\left\{j2\pi \frac{nd}{N}\right\}, \tag{2.96}$$

with a magnitude according to

$$|\iota_{N,\text{norm}}[d]| = \frac{r_0 + d \cdot r_\Delta}{r_0} \cdot \frac{1}{N} \sum_{n=0}^{N-1} \exp\left\{ \text{j} 2\pi \frac{nd}{N} \right\}. \tag{2.97}$$

when normalized to ρ. This result is obviously periodic with period N, resulting in an *unambiguous range*

$$r_\text{u} = N \cdot r_\Delta, \tag{2.98}$$

where r_Δ is the width of a range bin, which is

$$r_\Delta = \frac{c}{2N f_\Delta} = \frac{c}{2B} \tag{2.99}$$

according to (2.95), with B being the width of the spectrum covered by the transmitted signals. The width of a range bin is related to the range-resolution, and it is equal to expression (2.89). However, it does not correspond to the FWHM of the processing result, which has been used to define resolutions in the preceding sections. To find the FWHM range-resolution, (2.97) has to be evaluated for positions corresponding to non-integer values of d. The processing result is the Dirichlet kernel [46], multiplied by a range dependent term. Around its maximum, the Dirichlet kernel has an approximately sinc-like shape, with the first nulls around r_0 separated by $2r_\Delta$ and a corresponding FWHM of

$$\delta_r = 0.6 \cdot 2r_\Delta = 0.6 \cdot \frac{c}{B} \tag{2.100}$$

Figure 2.14 shows the magnitudes of processing results normalized to ρ for a target at range $r_0 = 1\,\text{m}$, for $\alpha_0 = 10°$. The plot in fig. 2.14(a) has been generated for 1, 3, and 21 frequency points spaced by $f_\Delta = 100\,\text{MHz}$ and centered at $f_c = 24\,\text{GHz}$. The periodic contribution of the IDFT is clearly visible. Processing the data for a target generates one peak at the location of the target and additional peaks spaced by r_u, which for the given setup is $1.5\,\text{m}$ according to (2.98). As N is increased, the magnitudes of those peaks remain constant. The level of the envelope of the Dirichlet kernel, however, decreases, which means that the effect of masking other targets is reduced. The plot in fig. 2.14(a) has been generated for a constant number $N = 21$ of frequencies centered at $24\,\text{GHz}$, but with varying f_Δ. As f_Δ is decreased, the covered bandwidth decreases and δ_r increases. At the same time, the unambiguous range, i.e. the spacing between the peaks, is increased. It is clearly visible that the peak spacing doubles as f_Δ is reduced from $100\,\text{MHz}$ (dotted) to $50\,\text{MHz}$ (dash-dotted), and again as it is reduced from $50\,\text{MHz}$ to $25\,\text{MHz}$ (solid). The plots reveal that $|\iota|$ is not symmetric about $r_0 + \Delta r = r_0$. The envelope of the peaks increases with Δr as given by (2.87) and (2.88). However, as N is increased, the magnitude of the envelope of the Dirichlet kernel is reduced between the peaks, and the effect of masking other targets is reduced. As with single-tone imaging, dividing the processing result by $r_0 + \Delta r$ would prevent the peaks' levels from rising with $r_0 + \Delta r$, but the effect of masking would remain the same since then targets at greater ranges would cause lower peaks. Weighting the cross-range processing results obtained for different frequencies by proper weighting functions instead of merely adding them

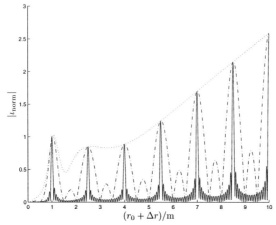

(a) $f_\Delta = 100\,\mathrm{MHz}$. Dotted: $N = 1$, dash-dotted: $N = 3$, solid: $N = 21$

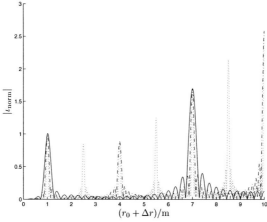

(b) $N = 21$. Solid: $f_\Delta = 25\,\mathrm{MHz}$, dash-dotted: $f_\Delta = 50\,\mathrm{MHz}$, dotted: $f_\Delta = 100\,\mathrm{MHz}$

Figure 2.14: Normalized processing results for varying f_Δ and N for $r_0 = 1\,\mathrm{m}$ and $f_\mathrm{c} = 24\,\mathrm{GHz}$

with equal weights, the side-lobe levels of the Dirichlet kernels can be reduced. For a given imaging task, N and f_Δ and potentially a weighting function have to be chosen such that the unambiguous range is great enough, and that the requirements regarding side-lobe levels within the unambiguous range and the range-resolution are met.

For large integration angles and large signal bandwidths at the same time, both range and cross-range resolution depend on both the bandwidth of the transmitted signal and the region of aspect angles, for which the target is observed and observable. In [42], this fact is recognized and accounted for by discussing a resolution area rather than resolution expressions for range and cross-range directions separately. The resolution area—with resolution being defined based on the equivalent rectangle width under the prerequisite of sufficiently large integrated mainlobe-to-sidelobe energy ratio (ISLR)—is determined to be

$$A_{\min} = \frac{\lambda_{\mathrm{c}}}{2\Delta\vartheta} \frac{c}{2B}, \tag{2.101}$$

where λ_{c} is the wavelength corresponding to the frequency at the center of the bandwidth, and $\Delta\vartheta$ is the total beam-width of the antenna. For very small signal bandwidths, the ISLR might be poor and thus (2.101) might be invalid. According to [42], an ISLR greater than $3\,\mathrm{dB}$ is obtained whenever the bandwidth of the transmitted signal is chosen to be

$$B > f_{\max}\left[1 - \cos(\Delta\vartheta/2)\right]. \tag{2.102}$$

The contour plot in fig. 2.13 reveals that side-lobes higher than $-6\,\mathrm{dB}$ with respect to the maximum are present around the location of the directive target that can be diminished by increasing the bandwidth of the transmitted signal above zero.

2.6 Sampling requirements

Before processing, the signal observable along the synthetic aperture is sampled at discrete locations. For an alias-free reconstruction, the Nyquist criterion has to be fulfilled, which means that the signal has to be sampled at a rate at least twice as high as the one-sided bandwidth occupied by the signal. As shown in section 2.5.1, the spatial bandwidth occupied by the received signal (2.27) due to an omni-directional target observed within $\pm\alpha_0$ is

$$\Delta_{k_x} = 4k\sin\alpha_0$$

according to (2.62). The spectrum is symmetric about $k_x = 0$. Thus, the maximum allowable spacing between adjacent samples is

$$\Delta_{x,\max} = \frac{2\pi}{\Delta_{k_x}} = \frac{\lambda}{4\sin\alpha_0}, \tag{2.103}$$

which is related to the cross-range resolution by

$$\Delta_{x,\max} = \frac{\delta_x}{1.2} \tag{2.104}$$

according to (2.65). For narrow antenna beams and $a_{\mathrm{tf}} = 1$, the maximum allowable sample spacing can be approximated using (2.66) as

$$\Delta_{x,\max} = \frac{l}{2}, \tag{2.105}$$

where l is the extent of the real antenna in x-direction. This is what is commonly stated for narrow-beam SAR systems in the literature, e.g. in [39, p. 646]: The sample spacing does not have to exceed half the extent of the antenna. For wide antenna beams, $\Delta_{x,\text{max}}$ approaches $\lambda/4$. Figure 2.15 shows a plot of the maximum allowable sample spacing

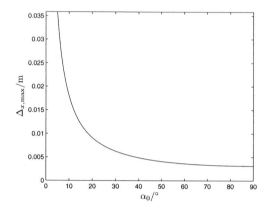

Figure 2.15: Maximum allowable sample spacing $\Delta_{x,\text{max}}$ vs. α_0

vs. α_0 for the wavelength $\lambda = 1.25\,\text{cm}$. As can be seen, the allowable sample spacing reduces by a factor of approximately 10 from several centimeters for $\alpha_0 = 5°$ to a few millimeters as α_0 approaches $90°$.

2.7 Simulations for selected types of targets

In this section, the processing results for selected types of (hypothetic) targets will be discussed, clarifying the influence of signal, antenna, and target properties. The plots show the magnitudes of the processing results in decibels, versus the cross-range coordinate x' and the range coordinate r', normalized to its value at the true location of the target. r' rather than e.g. y' is used as the range coordinate since the result of processing data acquired along a one-dimensional synthetic aperture is ambiguous with respect to β as considered in section 2.1. Processing is done according to the weighted compensation approach—as introduced in section 2.4—for each of the frequencies used, and then combining the intermediate results of each frequency in a range-stacking-like manner as discussed in section 2.5.3. For all the simulations, the center frequency is 24 GHz. The targets' reflectivities are assumed to be invariant with respect to the frequency. Variations of the reflectivity with the frequency would have effects similar to those which are encountered when the reflectivity is a function of the aspect angle.

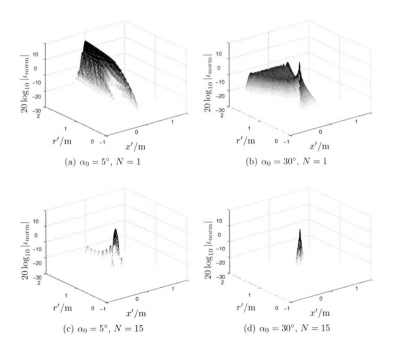

Figure 2.16: Omni-directional target at $x = 0$, $r = 1\,\mathrm{m}$, $f_\Delta = 80\,\mathrm{MHz}$

Figure 2.16 shows the processing results for an omni-directional target at $x = 0$ and $r = 1\,\mathrm{m}$ for different combinations of beam-widths and numbers of frequency points. For the comparatively small beam-width $2\alpha_0 = 10°$, the processing result does not even exhibit a local maximum at the position of the target, when only a single frequency is used (fig. 2.16(a)). Increasing the beam-width (fig. 2.16(b)) or the number of frequencies (fig. 2.16(c)) yields a local maximum at the location of the target, and a combination of both measures yields even lower side-lobes (fig. 2.16(d)).

The benefit of not only increasing the beam-width but additionally increasing the bandwidth of the transmitted signal (by increasing the number of frequency steps) becomes obvious from inspecting fig. 2.17. The setup is the same as the one discussed above, with the exception that the target is directive. It is assumed that its reflectivity is non-zero only for aspect angles α in the region $[-1°, 1°]$. Even with an antenna with a beam-width of 60° (fig. 2.17(a)), the range resolution is poor, and the processing result does not exhibit a local maximum at the location of the target. Increasing the bandwidth of the transmitted signal causes the processing result to show a maximum at the location of the target and to exhibit resolution in range direction according to the bandwidth of the transmitted signal (fig. 2.17(b)). The cross-range resolution is dictated by the limited observability of the target rather than the (large) beam-width of the antenna.

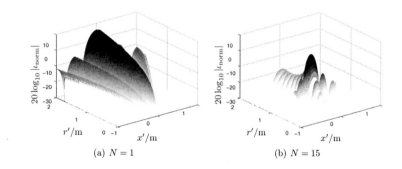

(a) $N = 1$　　　　　　　　　　　　　(b) $N = 15$

Figure 2.17: Directive target at $x = 0$, $r = 1\,\mathrm{m}$, observable for $\alpha \in [-1°, 1°]$, $f_\Delta = 80\,\mathrm{MHz}$, $\alpha_0 = 60°$

The influence on the processing result of applying a windowing function in the k_x-domain in order to control the cross-range side-lobe levels becomes obvious from fig. 2.18. The scene consists of two directive targets, located at $x = \pm0.5\,\mathrm{m}$ and $r = 1\,\mathrm{m}$. The targets have equal reflectivity characteristics, but they have different orientations. The targets' reflectivities are non-zero within a region of aspect angles having an extent of 10°. The target at $x = 0.5\,\mathrm{m}$ is oriented such that its reflectivity pattern's axis of symmetry is perpendicular to the x-axis, whereas the target at $x = -0.5\,\mathrm{m}$ is rotated by 20°. Such a target is called a *squint* target. The half beam-width of the antenna is $\alpha_0 = 30°$. Applying a Hann window in the k_x-domain causes the reflectivity contributions from different angles to be weighted non-uniformly as discussed in section 2.5.1. Contributions from the edges of the angular support of the antenna pattern are weighted less than those from the center of the pattern. Consequently, the squinted target appears 13 dB weaker than the unsquinted target (fig. 2.18(a)). Rendering sidelobe control in favor of uniform weighting by applying a rectangular window results in peaks of equal heights (fig. 2.18(b)). Figure 2.19 shows a top view of the processing

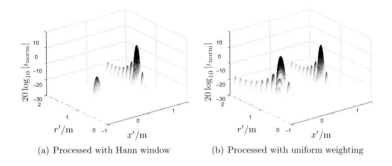

(a) Processed with Hann window   (b) Processed with uniform weighting

Figure 2.18: Squinted and unsquinted directive targets

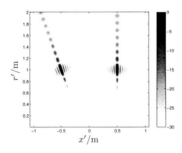

Figure 2.19: Top view of fig. 2.18(b)

result for uniform reflectivity weighting. The system response to the squinted target is very similar to that to the unsquinted one, except that it is rotated by the same angle as the target is rotated. This property is documented for matched-filter processing in [35, p. 292].

The lack of range resolution, when strongly directive targets are imaged using a single frequency, as previously discussed for a point-like target, is illustrated in fig. 2.20 for an extensive target. The scene is assumed to consist of (an idealized model of) a metal plate of width $1\,\mathrm{m}$ centered at $(0, 1\,\mathrm{m}, 0)$, which is perpendicular to the x-y-plane and whose surface normal intersects the x-z-plane with an angle of $20°$. The received signal is assumed to be produced by purely specular reflexions. The single-tone processing result for that scene is shown in fig. 2.20(a) and fig. 2.20(b). Using 15 frequencies spaced by $80\,\mathrm{MHz}$ instead of only a single one, the corresponding processing result shown in fig. 2.20(c) and fig. 2.20(d) represents the imaged scene much better, showing the approximate dimension, the location and the orientation of the target, along with side-lobes in the direction typical for a squint target.

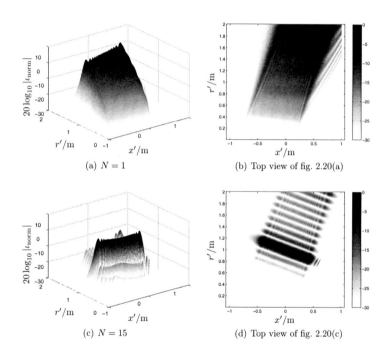

(a) $N = 1$ (b) Top view of fig. 2.20(a)

(c) $N = 15$ (d) Top view of fig. 2.20(c)

Figure 2.20: Model of squinted metal plate, $\alpha_0 = 30°$

Real world targets might exhibit reflectivities with not only their magnitude but also their phase varying over the aspect angle. Such targets are called range extended targets in [35], since their processing results might be more extended than the processing results of ideal point-targets. Furthermore, the processing result at the location of the target might be zero despite the presence of the target. This can occur when then mean of the real part of the reflectivity and the mean of the imaginary part of the reflectivity both are equal to zero. Figure 2.21 shows a comparison of the processing results for such a hypothetic zero-mean target and a target with constant reflectivity, both being imaged using an antenna with $\alpha_0 = 30°$ and one single frequency. The single-tone results in fig. 2.21 for the non-zero mean (fig. 2.21(a)) and the zero mean target (fig. 2.21(c)) look equal at first glance. However, comparing the corresponding top views in fig. 2.21(b) and fig. 2.21(d), respectively, reveals that the processing result of the zero-mean target is zero at the position of the target. Instead, the processing result exhibits a peak of almost the same height close to the true location of the target.

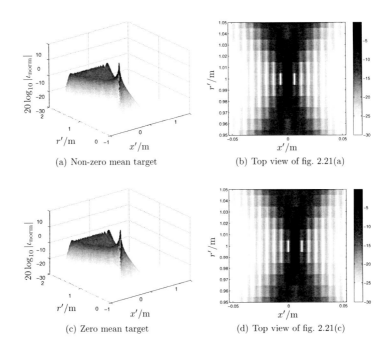

(a) Non-zero mean target

(b) Top view of fig. 2.21(a)

(c) Zero mean target

(d) Top view of fig. 2.21(c)

Figure 2.21: Non-zero mean and zero mean targets, located at $x = 0$, $r = 1\,\text{m}$, imaged with $\alpha_0 = 30°$, $N = 1$

Virtually the same result as for a point-target is obtained when the target is assumed to be a metallic sphere with a radius that is much smaller than the range of the target (fig. 2.22). In this case, the signals received along the synthetic aperture have amplitudes that are quasi equal to that for a target with zero radius, but their phase differs by a constant amount dictated by the radius of the sphere and the wavelength of the radiated signal, since the reflexion takes place at the sphere's surface. As a result, the peak in the SAR image occurs at the very same position as that for a zero-radius target. It only differs in phase, while the magnitude is virtually equal.

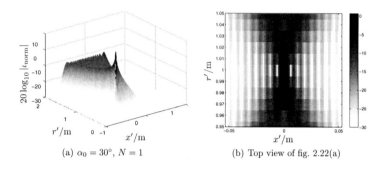

(a) $\alpha_0 = 30°$, $N = 1$ (b) Top view of fig. 2.22(a)

Figure 2.22: Metallic sphere with radius $4\,\mathrm{cm}$ located at $x = 0$, $r = 1\,\mathrm{m}$

2.8 Quality of the estimation

In the previous section, system responses for different kinds of targets and processing setups have been investigated. It was shown that the information conveyed in the processing result depends on properties of both the target and the system. Targets might be unobservable, e.g. if they are oriented in such a way that the incident wave is reflected into directions for which the system is not sensitive. This would happen e.g. for a metal plate, whose processing result is depicted in fig. 2.20, if it was imaged with an antenna with α_0 being smaller than the angle of rotation of the target. A comparable effect is observed with weighted processing, where the strength of the processing result for one and the same target depends on its orientation as discussed with fig. 2.18.

It is the aim of this section to provide a measure for the quality of the imaging result regarding the significance of its value at the position of the target, and to discuss ways to optimize the quality. Two effects will be considered that might disturb the processing result. The first one is thermal noise as it is present in every real-world receiver. The signal-to-noise ratio (SNR) is a familiar figure of merit in signal processing. To fit into the SNR framework, and as suggested for a class of similar effects in [47, p. 18.34 ff], the second effect will be referred to as *target noise*, describing the mean estimation error due to the limited region of aspect angles determined by the beam-width of the antenna. In [48], the target noise contribution has been evaluated for a single class of targets with a number of specular reflexions, disregarding the influence of thermal noise. As in [49], both thermal and target noise contributions will be considered in the following.

In order to determine estimation errors, the value that is to be estimated has to be identified. What is commonly displayed and regarded as *the* SAR image is the spatial distribution of the mean power reflexion coefficients within the resolution cells, where the—potentially weighted—mean is taken over the different aspect angles supported by the antenna along its path, s. e.g. [24, p. 139]. It is commonly assumed for narrow-beam SAR systems that the reflectivity is almost constant within the region of aspect angles and therefore, the mean power backscatter coefficient, $\bar{\sigma}$, is equal to the squared magnitude of the mean amplitude reflexion coefficient, $\bar{\rho}$,

$$\bar{\sigma} = |\bar{\rho}|^2. \tag{2.106}$$

For targets with a reflectivity that varies with the aspect angle, (2.106) is not valid. This is easily verified for a hypothetic target with an amplitude reflectivity that takes the values $+1$ and -1 (disregarding units) for angular intervals of equal extents within the observed region, resulting in a mean amplitude reflexion coefficient

$$\bar{\rho} = 0, \tag{2.107}$$

but a mean power reflexion coefficient

$$\bar{\sigma} = 1. \tag{2.108}$$

To obtain correct estimates of $\bar{\sigma}$ from estimates of $\bar{\rho}$, the estimates of $\bar{\rho}$ have to be taken over small angular regions, so that ρ is constant within each of those regions, and $\bar{\sigma}$ has to be determined from the squared magnitudes of $\bar{\rho}$ obtained for different regions of aspect angles. In general, this would have to be done for infinitely small regions of aspect angles. Then, however, the formerly coherent system would turn into a non-coherent system, which is not capable of producing cross-range resolution superior to that given by the width of the area illuminated due to the beam-width of the antenna. Thus, SAR imaging generally yields estimates of $\bar{\rho}$ rather than $\bar{\sigma}$.

Allowing the reflectivity to vary over the aspect angle rather than requiring it to be constant, wide-beam SAR imaging can be interpreted as estimating the mean reflectivity of a target over the entirety of aspect angles, based on the observation within a region of aspect angles limited by the width of the antenna beam. For this interpretation, the value desired to be estimated is the mean amplitude reflectivity within the region $[-\pi, \pi]$ of aspect angles, i.e. from every possible direction within the plane defined by the synthetic aperture and the location of the target,

$$\bar{\rho}_\pi = \frac{1}{2\pi} \int\limits_{-\pi}^{\pi} \rho(\alpha) \mathrm{d}\alpha. \tag{2.109}$$

The real-world processing is subject to different noise influences, two of which are thermal noise and target noise.

Thermal noise is considered an additive process that superimposes a statistical signal to the signal expected to be received according to (2.27). Letting $n(t)$ be the complex thermal noise contribution, and with (2.36) and (2.26), the estimate of the mean reflectivity within $\pm\alpha_0$ of a target at cross-range coordinate $x = 0$ and range r_0 reads as

$$\hat{\bar{\rho}}_{\alpha_0} = \frac{1}{2\alpha_0 a} \int\limits_{-\alpha_0}^{\alpha_0} [s_\alpha(\alpha) + n(t)] \cdot \frac{r_\alpha^2(\alpha)}{D_A^2(\alpha)} \cdot \exp\{j2kr_\alpha(\alpha)\}\, d\alpha$$

$$= \bar{\rho}_{\alpha_0} + \frac{1}{2\alpha_0 a} \int\limits_{-\alpha_0}^{\alpha_0} n'(\alpha, t) d\alpha \tag{2.110}$$

with

$$n'(\alpha, t) = n(t) \cdot \frac{r_\alpha^2(\alpha)}{D_A^2(\alpha)} \cdot \exp\{j2kr_\alpha(\alpha)\} \tag{2.111}$$

being the thermal noise contribution weighted in the processing algorithm.

Using this estimate obtained for a limited region of aspect angles as an estimate for $\bar{\rho}_\pi$, the estimation error ε is given by

$$\varepsilon = \bar{\rho}_\pi - \hat{\bar{\rho}}_\pi = \bar{\rho}_\pi - \hat{\bar{\rho}}_{\alpha_0} = \varepsilon_t - \varepsilon_r. \tag{2.112}$$

$$\varepsilon_t(\alpha_r, \alpha_0) = \bar{\rho}_\pi - \frac{1}{2\alpha_0} \int\limits_{-\alpha_0}^{\alpha_0} \rho(\alpha - \alpha_r) d\alpha \tag{2.113}$$

is the estimation error that results from inferring from $\bar{\rho}_{\alpha_0}$ to $\bar{\rho}_\pi$ and will be referred to as caused by target noise.

$$\varepsilon_r = \frac{1}{2\alpha_0 a} \int\limits_{-\alpha_0}^{\alpha_0} n'(\alpha, t) d\alpha \tag{2.114}$$

is the estimation error due to thermal receiver noise.

As a figure of merit for the quality of the estimation, the signal-to-thermal-and-target-noise ratio (SNR_{rt}) will be used, defined as

$$\text{SNR}_{rt} = \frac{S}{N}, \tag{2.115}$$

where

$$S = |\bar{\rho}_\pi|^2 \tag{2.116}$$

is the power of the signal to be estimated and

$$N = E\{|\varepsilon|^2\} = E\{\Re^2\{\varepsilon\}\} + E\{\Im^2\{\varepsilon\}\} \tag{2.117}$$

the power of the noise contribution. The subscript r in SNR_{rt} denotes thermal noise as it originates from the *r*eceiver, and t denotes *t*arget noise. $E\{\cdot\}$ denotes the expected

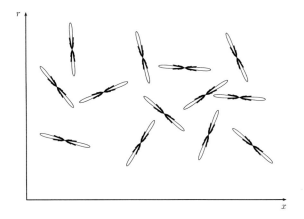

Figure 2.23: Scene consisting of a number of equal targets with varying orientation

value operator with respect to both time and orientation of the object. N is to be understood as the mean noise contribution over all possible orientations of the target and all realizations of the receiver noise process. Figure 2.23 shows a scene consisting of a number of targets with equal reflectivity patterns but different orientation. The SNR_{rt} is a measure for the ratio of the mean signal energy, taken over all pixels corresponding to the locations of an infinite number of randomly oriented (point) targets, to the mean squared estimation error due to target and receiver noise, taken over all targets and all realizations of the thermal noise process. Note that for subsequent measurements of the same scene, the receiver noise contributions will be different, and averaging several measurements will result in higher SNR_{rt}. For constant orientation of the targets, however, the target noise contribution will be exactly the same for all of those measurements, which means that averaging several measurements does not increase SNR_{rt} with respect to the target noise contribution.

Since the processing algorithm is an estimator for the mean reflectivity, ε is zero-mean for zero-mean receiver noise, and therefore, the power of the noise contribution can be written as

$$N = N_{\text{r}} + N_{\text{t}} \tag{2.118}$$

with

$$N_{\text{r}} = \text{Var}\left\{\Re\left\{\varepsilon_{\text{r}}\right\}\right\} + \text{Var}\left\{\Im\left\{\varepsilon_{\text{r}}\right\}\right\} \tag{2.119}$$

being the receiver noise contribution and

$$N_{\text{t}} = \text{Var}\left\{\Re\left\{\varepsilon_{\text{t}}\right\}\right\} + \text{Var}\left\{\Im\left\{\varepsilon_{\text{t}}\right\}\right\} \tag{2.120}$$

the target noise contribution.

First, the receiver noise contribution will be investigated, starting with the variance of the real part of ε_r.

$$\mathrm{Var}\left\{\Re\left\{\varepsilon_r\right\}\right\} = \mathrm{Var}\left\{\Re\left\{\frac{1}{2\alpha_0 a}\int\limits_{-\alpha_0}^{\alpha_0} n'(\alpha,t)\mathrm{d}\alpha\right\}\right\}$$
$$= \left(\frac{r_0}{2\alpha_0 a}\right)^2 \mathrm{Var}\left\{\int\limits_{-x_0}^{x_0} \Re\left\{\frac{n(t)}{D(\alpha(x))}\cdot\exp\left\{\mathrm{j}\phi(x)\right\}\right\}\mathrm{d}x\right\}, \quad (2.121)$$

with

$$\phi(x) = 2kr_x(x) - \phi_D(\alpha(x)), \quad (2.122)$$

where ϕ_D and D are the phase and the magnitude of D_A^2, respectively. With $n_\Re$ and $n_\Im$ the real and imaginary parts of n, the integrand in (2.121) is

$$\Re\left\{\frac{n(t)}{D(\alpha(x))}\cdot\exp\left\{\mathrm{j}\phi(x)\right\}\right\} = \frac{n_\Re(t)\cos(\phi(x)) - n_\Im(t)\sin(\phi(x))}{D(\alpha(x))}. \quad (2.123)$$

Thus, the variance of the real part of ε_r can be written as

$$\mathrm{Var}\left\{\Re\left\{\varepsilon_r\right\}\right\} = \left(\frac{r_0}{2\alpha_0 a}\right)^2 \mathrm{Var}\left\{X\right\} \quad (2.124)$$

with

$$X = \int\limits_{-x_0}^{x_0} \frac{n_\Re(t)\cos(\phi(x)) - n_\Im(t)\sin(\phi(x))}{D(\alpha(x))}\mathrm{d}x. \quad (2.125)$$

For further evaluation, the integral is temporarily approximated by a sum, where x is discretized in steps of size Δx according to

$$x = l\cdot\Delta x, \quad (2.126)$$

and where

$$l_0 = x_0/\Delta x, \quad (2.127)$$

assumed integer, and $-l_0$ are the summation limits. The reason for this approximation is, that in the continuous integral two "samples" of the noise process are directly adjacent to each other and therefore correlated. In reality, samples are taken with non-zero intervals between them. Assuming that those intervals are great enough, which is satisfied in many cases in reality, two samples of the noise process can be treated as uncorrelated. Additionally, the noise processes in inphase and quadrature channel are assumed to be uncorrelated. Then,

$$
\begin{aligned}
&\operatorname{Var}\{X\} \\
&\approx \operatorname{Var}\left\{\sum_{l=-l_0}^{l_0} \frac{n_{\Re}(t)\cos(\phi(l\Delta x))}{D(\alpha(l\Delta x))}\Delta x - \sum_{l=-l_0}^{l_0} \frac{n_{\Im}(t)\sin(\phi(l\Delta x))}{D(\alpha(l\Delta x))}\Delta x\right\} \\
&= \Delta x^2\left[\operatorname{Var}\left\{n_{\Re}(t)\sum_{l=-l_0}^{l_0}\frac{\cos(\phi(l\Delta x))}{D(\alpha(l\Delta x))}\right\} + \operatorname{Var}\left\{n_{\Im}(t)\sum_{l=-l_0}^{l_0}\frac{\sin(\phi(l\Delta x))}{D(\alpha(l\Delta x))}\right\}\right] \\
&= \Delta x^2\left[\operatorname{Var}\{n_{\Re}(t)\}\sum_{l=-l_0}^{l_0}\left(\frac{\cos(\phi(l\Delta x))}{D(\alpha(l\Delta x))}\right)^2 + \operatorname{Var}\{n_{\Im}(t)\}\sum_{l=-l_0}^{l_0}\left(\frac{\sin(\phi(l\Delta x))}{D(\alpha(l\Delta x))}\right)^2\right].
\end{aligned}
\tag{2.128}
$$

Assuming that the powers of the noise contributions in inphase and quadrature channel are equal, and their sum is

$$
P_{\mathrm{n}}' = 2\operatorname{Var}\{n_{\Re}(t)\} = 2\operatorname{Var}\{n_{\Im}(t)\},
\tag{2.129}
$$

with P_{n}' being normalized to be dimensionless according to

$$
P_{\mathrm{n}}' = \frac{P_{\mathrm{n}}\cdot R}{u_0^2},
\tag{2.130}
$$

the variance of X reads as

$$
\begin{aligned}
&\operatorname{Var}\{X\} \\
&\approx \Delta x^2\cdot\frac{1}{2}\cdot P_{\mathrm{n}}'\cdot\sum_{l=-l_0}^{l_0}\frac{\cos^2(\phi(l\Delta x))+\sin^2(\phi(l\Delta x))}{D^2(\alpha(l\Delta x))} \\
&= \Delta x\cdot\frac{1}{2}\cdot P_{\mathrm{n}}'\cdot\sum_{l=-l_0}^{l_0}\frac{1}{D^2(\alpha(l\Delta x))}\cdot\Delta x,
\end{aligned}
\tag{2.131}
$$

and written with an integral approximately as

$$
\operatorname{Var}\{X\} \approx \Delta x\cdot\frac{1}{2}\cdot P_{\mathrm{n}}'\cdot\int_{-x_0}^{x_0}\frac{1}{D^2(\alpha(x))}\mathrm{d}x.
\tag{2.132}
$$

Evaluating the variance of the imaginary part of ε_{r} yields exactly the same result, so that the total thermal noise contribution is twice that of a single channel and therefore reads as

$$
N_{\mathrm{r}} = \left(\frac{r_0}{2\alpha_0 a}\right)^2\cdot\Delta x\cdot P_{\mathrm{n}}'\cdot\int_{-x_0}^{x_0}\frac{1}{D^2(\alpha(x))}\mathrm{d}x,
\tag{2.133}
$$

or—after substituting $\mathrm{d}x$ by $\mathrm{d}\alpha$, a by its definition (2.28), and denormalizing P'_{n} according to (2.130)—as

$$N_{\mathrm{r}} = \frac{(4\pi r_0)^3 \cdot \Delta x \cdot P_{\mathrm{n}}}{(2\alpha_0)^2 \cdot \epsilon^2 \cdot \lambda^2 \cdot g \cdot P_{\mathrm{TX}}} \cdot \int\limits_{-\alpha_0}^{\alpha_0} \frac{1}{D^2(\alpha)\cos^2(\alpha)} \mathrm{d}\alpha. \tag{2.134}$$

The second noise effect to be considered is target noise. Equation (2.120) can equally be written as

$$N_{\mathrm{t}} = \frac{1}{2\pi} \int\limits_{-\pi}^{\pi} |\varepsilon_{\mathrm{t}}(\alpha_{\mathrm{r}}, \alpha_0)|^2 \, \mathrm{d}\alpha_{\mathrm{r}}, \tag{2.135}$$

where α_{r} denotes the rotation of the target. N_{t} is the second moment of the absolute value of the estimation error due to reflectivity fluctuations, calculated as its mean value over all possible target rotations α_{r}. In general, a target can be not only rotated but tilted as well, so that the noise contribution would have to be determined over all possible tilt and rotation angles. Here, it is assumed that the tilt angle is fixed and only the rotation angle is randomized. No general statement on the behavior of N_{t} for increasing α_0 is possible. For the special case of an object with constant reflectivity, as for a metallic sphere, there is no target noise, irrespective of α_0. For targets with varying reflectivity, target noise may, but does not necessarily decrease monotonically with increasing α_0. N_{t} has to be evaluated for the type of target for which the imaging system is to be optimized.

With (2.115), (2.116), (2.118), (2.133), and (2.135), $\mathrm{SNR}_{\mathrm{rt}}$ reads as

$$\mathrm{SNR}_{\mathrm{rt}} = |\bar{p}_\pi|^2 \cdot \left[\frac{(4\pi r_0)^3 \Delta x P_{\mathrm{n}}}{(2\alpha_0)^2 \epsilon^2 \lambda^2 g P_{\mathrm{TX}}} \int\limits_{-\alpha_0}^{\alpha_0} \frac{1}{D^2(\alpha)\cos^2(\alpha)} \mathrm{d}\alpha + \frac{1}{2\pi} \int\limits_{-\pi}^{\pi} |\varepsilon_{\mathrm{t}}(\alpha_{\mathrm{r}}, \alpha_0)|^2 \, \mathrm{d}\alpha_{\mathrm{r}} \right]^{-1}. \tag{2.136}$$

For further discussion, it will be useful to define the signal-to-thermal-noise power ratio $\mathrm{SNR}_{\mathrm{r}}$, which equals the $\mathrm{SNR}_{\mathrm{rt}}$ for zero target noise contribution, i.e. for an omnidirectional target, and therefore reads as

$$\mathrm{SNR}_{\mathrm{r}} = |\bar{p}_\pi|^2 \cdot \left[\frac{(4\pi r_0)^3 \Delta x P_{\mathrm{n}}}{(2\alpha_0)^2 \epsilon^2 \lambda^2 g P_{\mathrm{TX}}} \int\limits_{-\alpha_0}^{\alpha_0} \frac{1}{D^2(\alpha)\cos^2(\alpha)} \mathrm{d}\alpha \right]^{-1}. \tag{2.137}$$

2.9 Optimum antenna patterns

Generally, a high reliability of the processing result with respect to its accuracy is desirable. In this section, it will be shown, how the freedom of choice of the antenna pattern can be used to maximize $\mathrm{SNR}_{\mathrm{rt}}$ for a given scenario. The beam-width of an antenna will be defined as that angular extent $2\alpha_0$, symmetric about $\alpha = 0$, in which the antenna pattern is non-zero. The antenna pattern is assumed to be zero outside the beam-width. Additionally, it is assumed that the integration as part of the processing uses all data observed within $\pm\alpha_0$, i.e. that the integration limits are $\pm\alpha_0$.

For a given target with unknown orientation, the only system parameter influencing the target noise contribution is the beam-width of the antenna. For fixed α_0, the target noise contribution is fixed. Then, the actual shape of the antenna pattern within $\pm\alpha_0$ can be used to maximize SNR_{rt}, which is achieved by minimizing the expression

$$\int\limits_{-\alpha_0}^{\alpha_0} \frac{1}{D^2(\alpha)\cos^2(\alpha)}\mathrm{d}\alpha \tag{2.138}$$

in (2.136) for otherwise constant system parameters. In general, the directivity, integrated over all solid angle elements of a sphere, is equal to 4π [50], i.e.

$$\int\int D(\Omega)\mathrm{d}\Omega = 4\pi, \tag{2.139}$$

and particularly for the orientation of the coordinate system as discussed in section 2.1,

$$\int\limits_{-\pi}^{\pi}\int\limits_{-\pi/2}^{\pi/2} D(\alpha,\beta)\cos(\alpha)\mathrm{d}\alpha\mathrm{d}\beta = 4\pi, \tag{2.140}$$

which is the boundary condition for the minimization of (2.138). The first step to maximizing SNR_{rt} for given target noise contribution is to restrict the radiation to only those regions of space, from where signals will be processed. For a given maximum processing angle α_0, the antenna pattern has to vanish for $|\alpha| > \alpha_0$. For the following considerations, it is assumed that the antenna pattern does not vary with the angle β except for the fact that the pattern vanishes for $|\beta| > \beta_0$, resulting in a limited field of view, within which the SNR_{rt} for an omni-directional target is independent of its location with respect to the related angle β. Under the prerequisites of a pattern vanishing for $|\alpha| > \alpha_0$ and $|\beta| > \beta_0$, and being constant with respect to β inside its support region, the boundary condition can be written as

$$2\beta_0 \int\limits_{-\alpha_0}^{\alpha_0} D(\alpha)\cos(\alpha)\mathrm{d}\alpha = 4\pi. \tag{2.141}$$

The optimization can be performed by means of Lagrange multipliers and the calculus of variations [51]. With the substitution

$$e(\alpha) = D(\alpha)\cos(\alpha), \tag{2.142}$$

the Lagrange functional

$$
\begin{aligned}
F(\alpha) &= \int_{-\alpha_0}^{\alpha_0} \left(\frac{1}{e^2(\alpha)} + \gamma e(\alpha) \right) d\alpha - \gamma \frac{2\pi}{\beta_0} \\
&= \int_{-\alpha_0}^{\alpha_0} f(e(\alpha)) d\alpha - \gamma \frac{2\pi}{\beta_0}
\end{aligned}
\tag{2.143}
$$

with

$$
f(e(\alpha)) = \frac{1}{e^2(\alpha)} + \gamma e(\alpha)
\tag{2.144}
$$

can be formed. A necessary condition for D being an extremal is

$$
f_e = 0.
\tag{2.145}
$$

This condition is fulfilled by

$$
D(\alpha) = \left(\frac{2}{\gamma} \right)^{\frac{1}{3}} \cdot \frac{1}{\cos(\alpha)}.
\tag{2.146}
$$

Using the boundary condition (2.141), γ can be determined to be

$$
\gamma = 2 \left(\frac{\beta_0 \alpha_0}{\pi} \right)^3.
\tag{2.147}
$$

With (2.146), the pattern that maximizes SNR_{rt} for given target noise contribution and a given value of α_0 is

$$
D_{opt,\alpha_0}(\alpha) = \left\{ \begin{array}{ll} \frac{\pi}{\alpha_0 \beta_0} \cdot \frac{1}{\cos(\alpha)} & \text{for } |\alpha| \leq \alpha_0 \text{ and } |\beta| \leq \beta_0 \\ 0 & \text{else} \end{array} \right. .
\tag{2.148}
$$

Since this pattern maximizes SNR_{rt} for given target noise contribution, it also does so for zero target noise contribution, and therefore, it also maximizes SNR_r for given α_0.

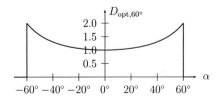

Figure 2.24: Pattern minimizing the influence of thermal noise for $\alpha_0 = 60°$

Using this optimum pattern, $\mathrm{SNR_{rt}}$ reads as

$$\mathrm{SNR}_{\mathrm{rt},D_{\mathrm{opt}},\alpha_0} = |\bar{\rho}_\pi|^2 \cdot \left[\frac{r_0^3 \cdot \Delta x \cdot \alpha_0 \cdot \beta_0^2}{2\pi^2} \cdot \frac{(4\pi)^3}{\lambda^2} \cdot \frac{P_\mathrm{n}}{P_\mathrm{TX} \cdot g} + \frac{1}{2\pi} \int_{-\pi}^{\pi} |\varepsilon_\mathrm{t}(\alpha_\mathrm{r}, \alpha_0)|^2 \, \mathrm{d}\alpha_\mathrm{r} \right]^{-1},$$

$$(2.149)$$

and $\mathrm{SNR_r}$ as

$$\mathrm{SNR}_{\mathrm{r},D_{\mathrm{opt}},\alpha_0} = |\bar{\rho}_\pi|^2 \cdot \left[\frac{r_0^3 \cdot \Delta x \cdot \alpha_0 \cdot \beta_0^2}{2\pi^2} \cdot \frac{(4\pi)^3}{\lambda^2} \cdot \frac{P_\mathrm{n}}{P_\mathrm{TX} \cdot g} \right]^{-1}. \qquad (2.150)$$

Still, apart from β_0, that is chosen according to the desired field of view, there is one degree of freedom in the optimum antenna pattern: α_0. When α_0 is required to have a certain value in order to achieve a predetermined resolution or to meet the sampling requirements along the synthetic aperture, it is not a degree of freedom. However, when not resolution, sampling requirements or the like pose constraints on α_0, it can be chosen so that $\mathrm{SNR_{rt}}$ is maximized. Evaluating (2.149) reveals that the influence of thermal noise increases as α_0 increases. Therefore, α_0 should be small. However, the behavior of the target noise term with α_0 depends on the target that is present in the scene. In order to find the antenna pattern that maximizes $\mathrm{SNR_{rt}}$, (2.149)—and especially its target noise term—has to be evaluated for the considered type of target, yielding the optimum value of α_0, which with (2.148) finally determines the optimum pattern. As an example, the shape of that pattern that minimizes the influence of thermal noise for a given $\alpha_0 = 60°$ is shown in fig. 2.24.

In the following subsections, the optimum patterns for a variety of imaging scenarios will be investigated, and the influence of using suboptimal patterns will be discussed.

2.9.1 Optimum patterns for metallic spheres and metallic discs as targets

In this subsection, (2.149) will be evaluated for two types of targets, that have also been investigated in [52] and that differ greatly in their target noise properties. The first one is a metallic sphere, the second one a metallic disc. The sphere is modeled to have an amplitude reflectivity

$$\rho_\mathrm{s}(\alpha) = \left(\pi r_\mathrm{s}^2 \right)^{\frac{1}{2}} = \sqrt{\pi} r_\mathrm{s}, \qquad (2.151)$$

where r_s is the radius of the sphere, and the amplitude reflectivity of the metallic disc is modeled as one specular reflexion on each side of the disc, i.e.

$$\rho_\mathrm{d}(\alpha) = \left(\frac{4\pi A_\mathrm{d}^2}{\lambda^2} \right)^{\frac{1}{2}} \cdot (\delta(\alpha - \alpha_\mathrm{d}) - \delta(\alpha - \alpha_\mathrm{d} + \pi)), \qquad (2.152)$$

where $A_\mathrm{d} = \pi r_\mathrm{d}^2$ is the area of the disc with radius r_d, and α_d gives the orientation of the disc. Figure 2.25 shows plots of the reflectivity functions according to (2.151) and

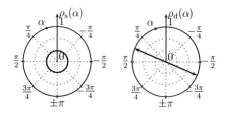

Figure 2.25: Reflectivities of metal sphere and metal disc

(2.152). The amplitudes of the reflectivity functions are taken as the square roots of the maximum radar cross sections stated in [39] for the considered objects. Choosing $r_s = 2r_d^2/\lambda$, the mean amplitude reflectivities over the whole circumference of both objects are equal. In contrast to that, the target noise contributions corresponding to their reflectivity functions differ greatly. The target noise contribution of the sphere is zero, $N_{t,s} = 0$, whereas evaluation of (2.135) reveals that the target noise contribution of the disc is

$$N_{t,d} = \frac{2A_d^2}{\lambda^2}\left(\frac{1}{\alpha_0} - \frac{2}{\pi}\right) \tag{2.153}$$

for $\alpha_0 \in [0, \pi/2]$. The limited area of validity of this expression is not a limitation, since the maximum aspect angle obtainable when the system is moved along a straight line is $\pi/2$. The target noise of the disc decreases for increasing α_0 and vanishes for $\alpha_0 = \pi/2$.

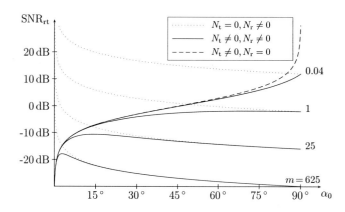

Figure 2.26: SNR_{rt} for varying system and object parameters

56

Fig. 2.26 shows the SNR_{rt} for different cases. The dashed curve corresponds to the signal-to-noise ratio obtained for the metal disc when no thermal noise is present. Due to the irregular nature of the target, the mean square estimation error is zero only for $\alpha_0 = \pi/2$.

The dotted curves correspond to the metal sphere with the same mean amplitude reflectivity as the disc in the presence of thermal noise, where the parameter m determining the vertical shift of the (logarithmically plotted) curve is given by

$$m = \frac{8\pi^2 \cdot r_0^3 \cdot \Delta x \cdot \beta_0^2 \cdot P_{\text{n}}}{A_{\text{d}}^2 \cdot P_{\text{TX}} \cdot g}. \tag{2.154}$$

Due to the absence of target noise in case of the sphere, the resulting SNR_{rt} is only affected by the thermal noise contribution, and therefore it is inversely proportional to α_0 according to (2.150).

The solid curves correspond to the metallic disc for varying m, i.e. different levels of thermal noise with respect to the level of target noise. As an example, a measurement setup characterized by $\Delta x = 3\,\text{mm}$, $\beta_0 = 10°$, $P_{\text{n}} = -70\,\text{dBm}$, $P_{\text{TX}} = 0\,\text{dBm}$, $g = -20\,\text{dB}$, and $A_{\text{d}} = \pi r_{\text{d}}^2$ with $r_{\text{d}} = 10\,\text{mm}$ the radius of the metallic disc at a distance $r_0 = 2.4\,\text{m}$ from the x-axis is considered. This choice of parameters yields $m = 1$. The corresponding curve exhibits its maximum at $\alpha_0 \approx 71°$. That means that for this type of target in the given distance and for the given system parameters, the mean squared error over all possible α_{r} and all possible realizations of the thermal noise process is minimized for $\alpha_0 \approx 71°$. For smaller values of m due to e.g. higher transmitted power or smaller object distance, the maximum SNR_{rt} is reached for greater α_0 since the estimation error is mainly caused by the reflectivity fluctuations of the target, whereas greater values of m account for stronger thermal noise influence, suggesting to choose α_0 smaller.

The m-values of the plotted curves differ by a factor of 25 corresponding to e.g. transmitted powers differing by 14 dB or object distances differing by a factor of 2.9.

The plot reveals that the optimum beam-width strongly depends on the properties and the distance of the object as well as on system parameters. Therefore, the optimum beam-width has to be determined for each setup at hand individually.

2.9.2 Optimum patterns for sub-aperture processing

As mentioned in section 2.8, the mean amplitude reflectivity of a given target might be zero, while its mean power reflectivity is non-zero. Coherently processing the data as discussed up to now might result in a zero-valued pixel at the location of the target as shown in section 2.7. One way to avoid this problem is to implement a bank of filters for different orientations of the particular target that is to be searched for in the present scene. In case of unknown orientation of the target, the optimum pattern is the one given by (2.148). This kind of processing additionally yields information on the orientation of the target, but at the expense of having to use a bank of filters and related computational load. Another way to avoid such a zero-valued pixel despite the presence of a target—while a potential degradation of the resolution for other targets in the scene has to be accepted—is dividing the synthetic aperture into several sub-apertures, coherently processing the data obtained for every sub-aperture separately, and combining the different processing results non-coherently by adding their squared magnitudes.

Modeling a point-target as exhibiting a reflectivity which varies with the aspect angle is somewhat over-idealizing, since a certain extent of the target is needed to produce a directivity for finite frequency. However, several omni-directional point-targets that are located within a resolution cell of the SAR image may result in an overall reflectivity that varies with the aspect angle. As the number of such point-targets per resolution cell becomes very high—which might be related to a large resolution cell—the overall reflectivity within that resolution cell is best described by statistical means. Processing two sets of data sampled at slightly different positions along the synthetic aperture may result in two strongly differing SAR images. The fluctuations between the images are called *speckle* [24, 25]. To obtain good estimates of the mean power reflectivity from estimates of the mean amplitude reflectivity—as they are naturally delivered by a coherent SAR system as discussed in section 2.4—so called *sub-aperture* or *multi-look* processing is often performed in order to reduce the speckle. Different kinds of sub-aperture processing exist, some of which use overlapping sub-apertures to use the available signal power as efficiently as possible. An overview of existing techniques can be found in [53]. That antenna pattern that minimizes the influence of thermal noise depends on the way sub-aperture processing is actually implemented. For the straightforward case of dividing the synthetic aperture into a number of apertures with equal k_x-bandwidths and using uniform k_x-weighting within each of the sub-apertures, the pattern that yields minimum and equal thermal noise contributions for each of the sub-apertures is easily derived as follows. As an example, the synthetic aperture with an extent determined by $\alpha_0 = 30°$ is assumed to be divided into 6 sub-apertures. For the sake of simplicity, dividing the aperture in chunks of equal angular extent is assumed to yield chunks of equal k_x-bandwidths, which is a fair approximation for the sufficiently small choice of α_0, since k_x and α are related via

$$k_x = -2k \sin \alpha \qquad (2.155)$$

as discussed in section 2.5.1, being almost linear in the considered range of angles. For each of the 10°-sub-apertures, the respective part of the antenna pattern is optimized independently in the way discussed at the beginning of section 2.9, resulting in individual disjoint parts of the pattern being proportional to $1/\cos(\alpha)$. The relative magnitudes of the parts of the pattern are then determined by equalizing the thermal noise contributions within the individual sub-apertures according to (2.138), where the limits of integration have to be chosen according to the limits of the sub-aperture under consideration. It turns out, that the relative magnitudes have to be chosen equal, so that the resulting optimum pattern in this case is exactly the same as the one given in (2.148). For considerably larger α_0, dividing the synthetic aperture in chunks of equal k_x-bandwidths results in chunks of unequal angular extent. The shapes of the sub-patterns minimizing the thermal noise contribution will remain the same, but the relative weights of the sub-patterns will no longer be equal, but have to be adjusted to keep the noise contributions in each of the sub-apertures equal.

Note again, that sub-aperture processing decreases the effective k_x bandwidth and therefore reduces the resolution obtainable for targets with constant reflectivity. For the case of large numbers of scatterers within single resolution cells, however, sub-aperture processing might be inevitable to obtain meaningful SAR images.

2.9.3 Influence of suboptimal patterns

By means of (2.148), that antenna pattern, that minimizes the influence of thermal noise on the processing result at the location of the target for the task of unweighted mean reflectivity estimation within $\pm\alpha_0$ can be determined. In practice, antenna patterns, even if designed exclusively for a given application, will deviate from the optimum pattern. In this subsection, the influence of using sub-optimal antenna patterns will be judged in terms of loss in SNR_r for a number of given antenna patterns. The loss in SNR_r will be determined by comparing—for varying α_0—the SNR_r obtained with the antenna to be judged with the one obtainable with the optimum pattern for the current value of α_0. The pattern under consideration will be evaluated for $\beta = 0$. For a meaningful comparison, the optimum patterns—with their shapes given by (2.148)—are scaled by a factor that guarantees that the integrated directivities in the considered β-slice are equal for the optimum pattern and the pattern to be judged. The plotted values of SNR_r will be normalized to that SNR_r obtained with the optimum pattern for $\alpha_0 = 90°$.

The first pattern to be judged is a pattern with constant directivity with respect to α within $\pm 90°$ as plotted with a solid line in fig. 2.27(a). The nominal half beam-

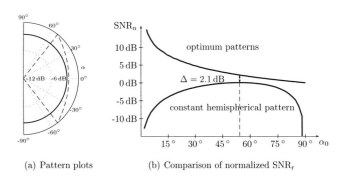

(a) Pattern plots   (b) Comparison of normalized SNR_r

Figure 2.27: Hemispherical pattern, optimum pattern and normalized SNR_r

width of such an antenna is $90°$. However, the plot in fig. 2.27(b) reveals that not $\alpha_0 = 90°$ but $\alpha_0 = 54°$ is that processing limit that yields maximum SNR_r. For small

α_0, the major part of the energy is radiated into directions that are not considered in the processing, and therefore the respective optimum pattern, which concentrates the energy to the considered directions only, yields much higher SNR_r. For $\alpha_0 = 54°$, the SNR_r obtainable with the pattern under consideration is maximum and only $2.1\,dB$ below the one achievable with the optimum pattern. The SNR_r maximizing pattern for $\alpha_0 = 54°$ is plotted dashed in fig. 2.27(a).

The pattern plotted with a solid line in fig. 2.28(a) is the one predicted for the horn antenna of section 3.2.3 according to the analytic horn antenna model in [50]. Evaluating the resulting SNR_r for varying α_0, it turns out that $\alpha_0 = 7°$ yields maximum SNR_r, $5.1\,dB$ below that of the respective optimum pattern plotted with a dashed line in fig. 2.28(a). For larger α_0, SNR_r drops relatively quickly.

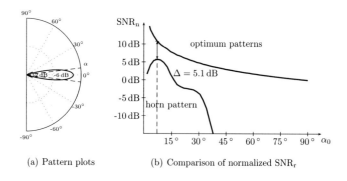

(a) Pattern plots   (b) Comparison of normalized SNR_r

Figure 2.28: Horn pattern, optimum pattern and normalized SNR_r

The SNR_r curve shown in fig. 2.29(b) corresponds to the pattern plotted with a solid line in fig. 2.29(a), which has been obtained by simulating the patch antenna structure discussed in section 3.2.3 with Agilent ADS for $\beta = 0$. The SNR_r-maximizing value of α_0 is $36°$ with a corresponding SNR_r difference of $3.2\,dB$ to the optimum antenna. The optimum antenna pattern for $\alpha_0 = 36°$ is plotted dashed in fig. 2.29(a). The patch array's SNR_r curve is comparatively flat over a wide range of α_0, allowing to vary α_0 without considerable changes in SNR_r.

The SNR_r plots and the procedure discussed above can be used to find that values of α_0 that minimize the influence of thermal noise on the processing result for given antenna patterns. In order to determine which one of two given antenna patterns is better suited for processing with a given value of α_0, the SNR_r has to be evaluated for the given absolute directivities rather than relative directivities as done above. A comparison in section 3.2.3 will show that even for such values of α_0, for which the SNR_r curve of the horn antenna has dropped by several decibels with respect to its maximum, the resulting SNR_r is still higher than that achievable with the patch array due to the horn's narrower beam-width with respect to β and the resulting higher

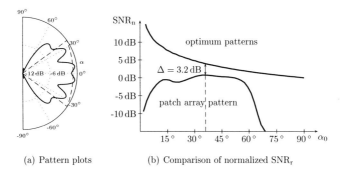

(a) Pattern plots (b) Comparison of normalized SNR_r

Figure 2.29: Patch array pattern, optimum pattern and normalized SNR_r

absolute directivity.

2.9.4 Influence of windowing

Up to now, only optimum patterns for *unweighted* mean reflectivity estimation have
been considered. For some applications, the use of windowing functions to control image
properties as discussed in section 2.5 might be desirable. As described there, windowing
in the k_x-domain has an inevitable influence on the weighting of reflectivity contributions from different directions. The weighting function in the α-domain is a non-linearly
scale-transformed version of the windowing function in the k_x-domain. In the following,
the influence of the weighting function on the SNR_r-maximizing antenna pattern will
be investigated. Figure 2.30 visualizes the influence of the weighting coefficients that

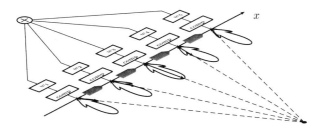

Figure 2.30: Illustration of the weighting process

affect the compensated signals received along the synthetic aperture—corresponding to varying aspect angles—before being summed. Considering this weighting, the signal-to-thermal-noise ratio can be determined following the same procedure as in section 2.8 to be—up to a constant of proportionality—

$$\text{SNR}_D^{w_2} = \frac{\left(\int\limits_{-\alpha_0}^{\alpha_0} w_{2,\alpha}(\alpha) \mathrm{d}\alpha \right)^2}{\int\limits_{-\alpha_0}^{\alpha_0} \frac{w_{2,\alpha}^2(\alpha)}{D^2(\alpha)\cos^2(\alpha)} \mathrm{d}\alpha} \tag{2.156}$$

for a target with constant reflectivity, where the subscript indicates the antenna pattern used for data acquisition and the superscript the (real-valued, non-negative) windowing function used for processing. An explicit subscript r indicating *receiver* thermal noise, as used before, has been omitted to avoid an even lengthier notation. Note that for the formulation of the SNR as stated above, those constants of proportionality that are not essential for the following discussion have been disregarded. Therefore, the SNR as used in this subsection, is to be understood as a relative measure rather than an absolute one.

For the special case of a rectangular windowing function—causing the reflectivity contributions for different aspect angles to be weighted uniformly—the antenna pattern yielding maximum and equal SNR for all β within $\pm\beta_0$ has been identified in (2.148) to be

$$D_{\text{opt,uni},\alpha_0}(\alpha) = \frac{\pi}{\alpha_0\beta_0} \cdot \frac{1}{\cos(\alpha)}, \tag{2.157}$$

for $|\alpha| \leq \alpha_0$, $|\beta| \leq \beta_0$ and zero outside. The resulting (normalized) SNR reads as

$$\text{SNR}_{D_{\text{opt,uni},\alpha_0}}^{\text{rect}} = \frac{2\pi^2}{\alpha_0\beta_0^2}. \tag{2.158}$$

However, in conjunction with other than rectangular weighting functions, the antenna pattern given in (2.157) is suboptimal concerning the achievable SNR. The SNR degradation, emerging when data obtained with an antenna pattern optimized for uniform weighting is processed with a non-rectangular window instead of a rectangular window, is

$$\Delta^- = \frac{\text{SNR}_{D_{\text{opt,uni},\alpha_0}}^{w_2}}{\text{SNR}_{D_{\text{opt,uni},\alpha_0}}^{\text{rect}}} = \frac{1}{2\alpha_0} \cdot \frac{\left(\int\limits_{-\alpha_0}^{\alpha_0} w_{2,\alpha}(\alpha)\mathrm{d}\alpha \right)^2}{\int\limits_{-\alpha_0}^{\alpha_0} w_{2,\alpha}^2(\alpha)\mathrm{d}\alpha}, \tag{2.159}$$

which is the reciprocal of the equivalent noise bandwidth (ENBW, s. e.g. [38]) of $w_{2,\alpha}$, and which by means of the Cauchy-Schwarz inequality [54] can be shown to be less than or equal to 1. Since α and k_x are related by

$$k_x = -2k\sin\alpha$$

as discussed earlier, $w_{2,\alpha}$ and w_{2,k_x} are of equal shapes for small integration angles only, and also the intrinsic window is almost constant for small integration angles only. That means that $1/\Delta^-$, which is the ENBW of $w_{2,\alpha}$, is equal to the ENBW of the k_x Hann window and therefore directly to be read from tables as in [38] only in case of small integration angles. For large integration angles, (2.159) has to be evaluated explicitly.

The SNR degradation, that has to be accepted when using a non-rectangular windowing function to process data that has been obtained with an antenna pattern optimized for uniform weighting, can be counteracted by using an antenna pattern optimized for the weighting function to be applied. Therefore, the SNR-maximizing antenna patterns will be derived for given weighting functions.

For a given weighting function, the numerator in (2.156) has a fixed value. That means that the SNR can be maximized by minimizing the denominator,

$$\int\limits_{-\alpha_0}^{\alpha_0} \frac{w_{2,\alpha}^2(\alpha)}{D^2(\alpha)\cos^2(\alpha)}\,\mathrm{d}\alpha, \tag{2.160}$$

with respect to D. Clearly, high values of D would cause this expression to be small. However, since D is the antenna's directivity, whose integral over the whole sphere has to be equal to 4π, the minimization is constrained by

$$2\beta_0 \int\limits_{-\alpha_0}^{\alpha_0} D(\alpha)\cos(\alpha)\mathrm{d}\alpha = 4\pi, \tag{2.161}$$

which can be written this way, since D is assumed to be constant with respect to β within $\pm\beta_0$ and zero outside, as discussed for uniform weighting at the beginning of section 2.9. Following the same optimization scheme, the optimum antenna pattern for weighted processing is found to be

$$D_{\mathrm{opt},w_2,\alpha_0}(\alpha) = \frac{w_{2,\alpha}^{2/3}(\alpha)}{\cos(\alpha)} \cdot \frac{2\pi}{\beta_0 \cdot \int\limits_{-\alpha_0}^{\alpha_0} w_{2,\alpha}^{2/3}(\alpha)\mathrm{d}\alpha}, \tag{2.162}$$

which, again, is different for every choice of α_0.

Comparing the SNR for the optimum patterns according to (2.156), and the one obtained for a rectangular window and an antenna optimized for the rectangular window, yields

$$\Delta^+ = \frac{\mathrm{SNR}_{D_{\mathrm{opt},w_2,\alpha_0}}^{w_2}}{\mathrm{SNR}_{D_{\mathrm{opt},\mathrm{uni},\alpha_0}}^{\mathrm{rect}}} = \frac{2\alpha_0 \cdot \left(\int\limits_{-\alpha_0}^{\alpha_0} w_{2,\alpha}(\alpha)\mathrm{d}\alpha\right)^2}{\left(\int\limits_{-\alpha_0}^{\alpha_0} w_{2,\alpha}^{2/3}(\alpha)\mathrm{d}\alpha\right)^3}, \tag{2.163}$$

which can be shown using Hölder's inequality [54] to be greater than or equal to 1. Δ^+ gives the SNR improvement that is achieved when data obtained with a pattern optimized for non-rectangular weighting is processed with the respective optimum window

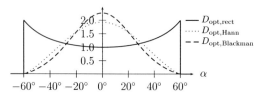

Figure 2.31: Shapes of optimum antenna patterns for various windowing functions and $\alpha_0 = 60°$

instead of when data obtained with a pattern optimized for rectangular weighting is processed with a rectangular window.

These results shall be discussed for three different weighting functions: an *uncompensated* rectangular window (unweighted mean reflectivity estimation, therefore side-lobe level varying over α_0 as discussed in section 2.5.1), a *compensated* Hann window (weighted combination of reflectivity contributions, first side-lobe: $-32\,\text{dB}$) and a *compensated* Blackman window (weighted combination of reflectivity contributions, first side-lobe: $-58\,\text{dB}$), where *compensated* means that the intrinsic window (2.64) has been considered according to (2.72) in order to achieve the desired properties of the Hann and the Blackman window, respectively.

The SNR-maximizing antenna patterns according to (2.162) for those three windows are shown in fig. 2.31 for $\alpha_0 = 60°$. A major difference between the patterns optimized for unweighted and windowed processing are the values towards $\alpha = \pm\alpha_0$, which play an important role when the actual processing is performed with another window than that for which the antenna pattern has been optimized, as will be discussed below. Figure 2.32 shows the SNR obtained with different combinations of windows and antenna patterns, normalized to the SNR obtained with a rectangular window and the corresponding optimum pattern, in dependence of α_0. Note that the SNR obtainable for uniform weighting, given by (2.158), decreases with α_0, and only the normalization causes its displayed values (e) to be constant.

Curves (f) and (g), representing the SNR obtained when data acquired with a pattern optimized for uniform weighting are processed with the compensated Hann and Blackman window, respectively, are located below (e), since $\Delta^- \leq 1$. Since $\Delta^+ \geq 1$, curves (a) and (c), representing Hann and Blackman weighting of data acquired with the respective optimum antenna patterns, are located above (e).

The curves (b) and (d) correspond to data acquired with a Hann optimized antenna pattern, processed with a Blackman window, and vice versa. The fact that curve (b) is located above (c)—which means that data acquired with a pattern optimized for the Hann window yields better SNR when processed with a Blackman window than with a Hann window—might be confusing, since a Hann window and a pattern optimized for a Hann window should be an optimal pair. Indeed, they are. The pattern optimized for a Hann window is that pattern that maximizes the SNR when data is to be processed with a Hann window. However, this does not necessarily mean that the Hann window is the

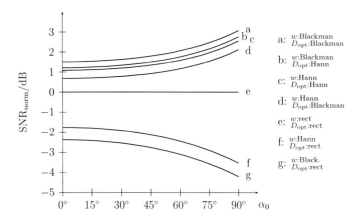

Figure 2.32: Normalized SNR obtainable with different combinations of antenna patterns and weighting functions

window that maximizes the SNR when used with an antenna optimized for the Hann window. Recall the optimization that has been carried out. Starting from a *given weighting function w* chosen for its properties (e.g. side-lobe levels), the numerator in (2.156) is fixed. A variation of D has influence on the denominator only. That means that for a given weighting function, the SNR is maximized by minimizing the denominator. However, in order to find the weighting function that maximizes the SNR for a *given antenna pattern*, the influence of the weighting function not only on the denominator, but also on the numerator has to be taken into account. Finding the pattern that maximizes the SNR for a given window is an optimization being different from finding the windowing function that maximizes the SNR for a given antenna pattern. Hence, the windowing function, for which the antenna pattern has been optimized, is not necessarily the windowing function that maximizes the SNR in conjunction with that antenna pattern.

Figure 2.32 does not show all possible combinations of the three considered windows and their corresponding optimum patterns. The reason is as follows. In reality, data are summed instead of integrated. For an antenna pattern with nulls (as is the case with the patterns optimized for the Hann and the Blackman windows) in conjunction with a windowing function that is non-zero at the respective position (e.g. the pattern optimized for the rectangular window), the numerical evaluation of the SNR according to (2.156) and therefore also the SNR obtained in a real world application depend strongly on the sample spacing and the actual positions of the samples. A variation of spacing and position can change the SNR actually achieved by several orders of magnitude. Therefore, plotting e.g. the SNR achievable when data acquired with a pattern optimized for a Hann window (including nulls) is processed with a rectangular window—giving a non-zero weight to the reciprocal of zero according to (2.156)—is not meaningful.

The findings of this section can be summed by stating that the optimum antenna pattern strongly depends on the imaging task and the processing actually to be implemented. Great benefit may arise from knowing the exact way of processing prior to specifying the antenna pattern. The investigations above can be used to judge given antennas for their applicability and to find optimum patterns for given imaging tasks.

2.10 Phase errors

In this section, a number of influences will be considered that alter the phase information—that one that is essential in forming a SAR image—conveyed in the signal that will be processed by the imaging algorithm. Phase errors can degrade the SAR image with respect to resolution, side-lobe levels, and the fidelity of the indicated target position, to name a few. The allowable error depends on requirements of a given imaging task. A phase error limit that is commonly regarded to be a conservative one is $\pi/8$. The phase errors discussed in the following subsections are due to undesired motion of the radar system, due to assuming a monostatic setup while using separate antennas for transmitting and receiving, and due to the gain imbalance of the quadrature mixer that is normally part of the system and will be described in section 3.2.2.

2.10.1 Motion errors

SAR imaging as described previously requires a linear path of the radar system along the scene. SAR systems for other than linear paths, such as circular ones (so-called circular SAR or CSAR [35]), are known and documented in the literature. What is common to all SAR systems is that the trajectory of the system has to be known. During data acquisition, a SAR system might deviate from its nominal path. In order to correct for the associated phase errors, the deviations have to be determined. The actual trajectory of the system can be found by external devices like global positioning system (GPS) receivers or acceleration sensors, or by analyzing the received signals for characteristics that indicate the system's trajectory [55]. Once the deviations are known, they can be corrected for by applying motion compensations algorithms (s. e.g. [56,57]). For small beam-widths, it might be sufficient to correct the sampled data for the phase error associated with the deviation. For wide-beam SAR imaging, it might be necessary to correct the signal contributions from targets at different aspect angles with angle dependent correction terms. Targets at different angles can be distinguished by evaluating the k_x-spectrum of the signals received along the synthetic aperture, since the instantaneous wave-number and the aspect angle are linked as discussed in section 2.5.1. An appropriate algorithm can be found e.g. in [58].

2.10.2 Phase error due to antenna separation

As stated in section 2.1, transmit and receive antenna are assumed to be co-located—the system is assumed to be monostatic. Whether this assumption is valid in a system that uses two antennas separated by a distance d can be judged on basis of the resulting phase error. This phase error originates from the fact that the electromagnetic wave, that is sensed by the receive (RX) antenna, has propagated from the transmit (TX) antenna to the target and from there to the receive antenna along a path of length $r_1 + r_2$, as depicted in fig. 2.33. Assuming a monostatic system, however, means that

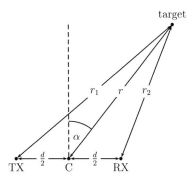

Figure 2.33: Geometry for determination of the phase error due to assumption of co-located antennas

one assumes the wave propagates from a point C, centered between both antennas, to the target and back, with a resulting path length $2r$. The difference between the actual and the assumed path length results in a phase error

$$\Delta\phi = \frac{2\pi}{\lambda}(r_1 + r_2 - 2r), \tag{2.164}$$

where λ is the wavelength of the transmitted signal. This phase error is maximum for $\alpha = 0$ and takes a value according to

$$\Delta\phi_{\text{max}} = \frac{4\pi}{\lambda}\left(\sqrt{r^2 + \left(\frac{d}{2}\right)^2} - r\right), \tag{2.165}$$

which increases with the antenna spacing d, and decreases with the range r of the target. Using (2.165), the maximum allowable antenna spacing can be determined for a target at a certain range for a given maximum phase error. Figure 2.34 shows a plot of the curve $d_{\text{max}}(r_{\text{min}})$ for a wavelength of 12.5 mm according to a frequency of 24 GHz for $\Delta\phi_{\text{max}} = \pi/8$.

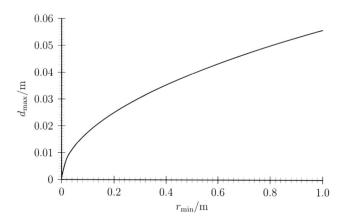

Figure 2.34: Maximum allowable antenna separation for given $\Delta\phi_{\max} = \pi/8$ at wavelength $\lambda = 12.5\,\mathrm{mm}$.

2.10.3 Phase error due to gain imbalance

Amplitude and phase of the received signal can be determined by means of an I/Q mixer. The inphase and quadrature branches potentially exhibit gain curves that vary with frequency. In case the variation is different in both branches, not only the amplitude, but also the phase—the crucial component for SAR imaging—is determined erroneous. In the following, the maximum allowable gain imbalance for a given phase error limit will be determined.

The received signal s is obtained from the inphase component s_I and the quadrature component s_Q as

$$s = s_\mathrm{I} - \mathrm{j}s_\mathrm{Q}. \qquad (2.166)$$

The phase of s is therefore

$$\phi_s = -\arctan\frac{s_\mathrm{Q}}{s_\mathrm{I}}, \qquad (2.167)$$

where the periodicity of the tangent function has not been considered for the sake of simplicity. The result of this investigation, however, remains the same. In case the inphase and quadrature branches exhibit amplitude gains g_I and g_Q, respectively, the determined phase is

$$\phi_{s,g} = -\arctan\frac{g_\mathrm{Q}s_\mathrm{Q}}{g_\mathrm{I}s_\mathrm{I}}. \qquad (2.168)$$

Defining

$$\sigma = \frac{s_\mathrm{Q}}{s_\mathrm{I}} \qquad (2.169)$$

and letting

$$\gamma = \frac{g_Q}{g_I}, \tag{2.170}$$

be the gain imbalance—where $g = 1$ means that the branches are balanced and $g \neq 1$ means they are not balanced—the phase error can be written as

$$\phi_{\mathrm{err}} = \phi_{s,g} - \phi_s = \arctan \frac{\sigma - \gamma\sigma}{1 + \gamma\sigma^2} \tag{2.171}$$

using trigonometric identities and assuming that g_I and g_Q have equal signs. This phase error function exhibits minima and maxima with equal magnitudes given by

$$\phi_{\mathrm{err,max}} = \left| \arctan \frac{1-\gamma}{2\sqrt{\gamma}} \right|. \tag{2.172}$$

Solving for γ, again using trigonometric identities, yields

$$\gamma_{1,2} = \frac{1 \mp \sin\phi_{\mathrm{err,max}}}{1 \pm \sin\phi_{\mathrm{err,max}}}. \tag{2.173}$$

γ_1 and γ_2 are reciprocals, which means that defining γ as its own reciprocal in (2.170) would have yielded exactly the same result. Therefore, the allowable range of gain imbalances for which the phase error is kept below a given maximum is

$$\left[\frac{1 - \sin\phi_{\mathrm{err,max}}}{1 + \sin\phi_{\mathrm{err,max}}}, \frac{1 + \sin\phi_{\mathrm{err,max}}}{1 - \sin\phi_{\mathrm{err,max}}} \right]. \tag{2.174}$$

As an example, a maximum tolerable phase error magnitude of $\pi/16$, i.e. a phase error difference of $\pi/8$ between the minimum and the maximum of the phase error function, is assumed. Then, the gains have to fulfill the condition

$$0.67 \leq \frac{g_Q}{g_I} \leq 1.48, \tag{2.175}$$

which is equal to

$$-3.4\,\mathrm{dB} \leq \frac{g_Q}{g_I} \leq 3.4\,\mathrm{dB} \tag{2.176}$$

on a logarithmic scale.

3 System implementation

In this chapter, the mechanical and electrical setup of a low-cost 23- to 25-GHz SFCW SAR system, successfully used for imaging tasks such as those reported on in the next chapter, will be described.

3.1 Mechanical setup

Figure 3.1: Mechanical setup of the implemented SAR system

In order to move the radar system on a linear path along the scene to be imaged, the components are mounted on a wooden vehicle, which is equipped with four wheels. The wheels exhibit v-grooves that guarantee that the vehicle can be moved on the rails of an ordinary ladder without deviations from the straight path. The vehicle holds all the equipment that is necessary to acquire and process radar data. The system is intended to be moved by hand. A rotary encoder, attached to the vehicle and equipped with a rubber wheel that rolls on top of one of the ladder's rails, is used to provide the system with information on its position. Figure 3.1 shows a photograph of the whole setup including all components.

3.2 Electrical setup

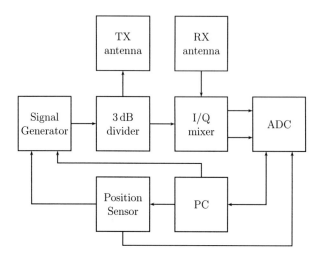

Figure 3.2: Block diagram of the SAR system

A block diagram of the SAR system is depicted in fig. 3.2. The system operates in SFCW mode [45], in which range resolution is decoupled from the instantaneous bandwidth—unlike with pulsed systems—and which therefore does not demand high-speed analog-to-digital converters (ADC). The radiofrequency (RF) part of the system comprises a signal generator, a power divider, transmit (TX) and receive (RX) antennas, and an I/Q mixer. The outputs of the mixer are connected to an ADC, which delivers its data to a personal computer (PC). Signal generator and ADC, as well as the position sensor used to trigger them, are configured by means of the PC. In the following subsections, each of those components and their interactions will be described in detail.

3.2.1 Signal generator

One of the key components of the SAR system is the signal generator. In-depth information is documented in [59]. The signal generator is controlled by a phase-locked loop (PLL), which can be configured by a PC via USB to generate CW signals at discrete frequency points in the range of 23 GHz to 25 GHz with an output power on the order of 20 dBm. As frequency points have to be stepped through as quickly as possible in order to allow fast motion of the vehicle, the frequencies to be generated are not adjusted one by one via USB, but a list of frequency points is loaded into the signal generator as part of the initialization of the radar system, and the frequency list is stepped through triggered directly by the position sensor. The settling time of the signal generator is below 20 μs as long as the frequency hop is not too large. In order to avoid the large hop from the upper end of the spectrum to the lower end, the frequencies are stepped through alternately in ascending and descending order. A key feature of the signal generator is its low cost. The signal generator consists of a variety of monolithic microwave integrated circuits (MMIC), including a 12 GHz voltage controlled oscillator (VCO), a frequency doubler, and an amplifier manufactured by Hittite Microwave Corporation with part numbers HMC515LP5, HMC448LC3B, and HMC442LC3B, respectively. The PLL (ADF4154, Analog Devices) operates around 3 GHz and is coupled to that output of the 12 GHz VCO that provides a copy of the actual output signal, with its frequency divided by 4. The commands issued by the PC via USB are preprocessed in a microcontroller (PIC18F4550, Microchip). The total price of all the mentioned integrated circuits is below 100 €. The photograph in fig. 3.3 shows the assembled components.

Figure 3.3: Configurable 23 GHz to 25 GHz signal generator with USB interface

3.2.2 I/Q mixer

Another key component of the SAR system is the I/Q mixer depicted in fig. 3.4. It comprises two double-balanced mixer MMICs with integrated local oscillator (LO) amplifiers for frequencies between 20 GHz and 43 GHz, manufactured by Agilent (part number HMMC-3040), and offered for approximately 20 € each. Figure 3.5 shows the layout of the microstrip lines that connect the signal and power supply ports of the MMICs to SMA connectors. The port marked with "RX" is connected to the RX antenna and delivers equal portions of the signal—via a Wilkinson power divider—to the RF ports of the MMICs. The port marked "Ref" is connected to the reference

Figure 3.4: I/Q mixer with opened housing

signal. The reference signal is split by means of another Wilkinson power divider and fed into the LO ports of the MMICs. The microstrip line towards the mixer on the left is designed to be a quarter-wavelength (at 24 GHz) longer than that to the mixer on the right, resulting in a phase difference of 90° between the reference signals fed into the mixers' LO ports. If necessary, the deviation of the phase difference from 90° due to the fixed difference in line-length for varying frequencies can be compensated by using the frequency dependent phase difference (rather than constantly 90°) for determining the phase of the received signal from the signals provided by the inphase (I) and quadrature (Q) outputs of the mixer.

The housing acts as a carrier for the MMICs, the substrate with the microstrip lines, and the SMA connectors, and as an electromagnetic shielding at the same time. The housing consists of two parts. The bottom part, along with the mounted components, is depicted in fig. 3.4. The interface between bottom part and top part is chosen such that it divides the drill-holes, that keep the SMA connectors for RF signals, into equal parts. For an adequate guidance of the RF signals, the dielectrics of the RF SMA connectors have to be tightly enclosed by the housing, i.e. the dielectrics have to be mounted into cylindric drill-holes with diameters equal to those of the dielectrics. Choosing the interface to be located above the drill holes would be a preferable solution, but the resulting comparatively high rim around the substrate would cause problems while bonding the wires that connect the microstrip lines and the MMICs. The chosen location of the interface circumvents bonding problems and allows to assemble the housing after electrically connecting the SMA connectors to the microstrip lines, while providing tightly enclosed dielectrics, which was not possible for an interface located below the center lines of the drill-holes.

The microstrip lines are gold plated, and the MMICs are connected to the microstrip lines by means of bond-wires. The substrate is punched at the positions marked with

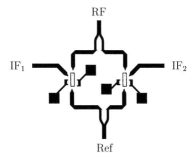

Figure 3.5: Layout of mixer's microstrip lines

empty rectangles in fig. 3.5. At the location of those cut-outs, the housing exhibits platforms of the same size with such a height, that the MMICs, when inserted into the cut-offs, exceed the top of the substrate by $100\,\mu$m to $200\,\mu$m, which is advantageous for bonding.

Four SMA connectors are used to supply each of the MMICs with positive and negative supply voltages separately. The use of SMA connectors even for the power supply allows the housing to be completely shielded. The supply lines are connected to conventional blocking capacitors, and to capacitances realized as patches on the substrate, that act as plate capacitors with values according to the MMICs' data sheet. An external power supply has been implemented that allows to control the mixers' gate voltages and therefore their operating points independently.

The I/Q mixer is fed with signals of equal frequencies at the RF and the reference ports, which allows to use a single signal generator rather than two, keeping the system costs low. As a consequence, the signals at the mixer's inphase and quadrature outputs have zero frequency, i.e. the output signals are DC signals for a given frequency and the vehicle standing still. From inphase and quadrature components, amplitude and phase of the signal received by the antenna can be retrieved according to (2.166) and (2.167).

A system using a single signal generator for generating the signal to be transmitted and the reference signal is called a homodyne system. A side-effect resulting from choosing RF and reference frequencies equal is the DC offset, that is superimposed on the inphase and quadrature signals due to the leakage of the reference signal (fed into the MMICs' LO ports) into the RF branch inside the MMICs. For the MMICs used, the LO-to-RF isolation is specified to be 18 dB. Calibration techniques to eliminate the DC offsets will be discussed in section 3.3.3.

3.2.3 Antennas

For the imaging experiments documented in the next chapter, two types of antennas have been used: horn antennas and patch antennas. Properties of those antennas will be discussed in the following subsections.

Horn antennas

Figure 3.6: TX and RX horn antennas mounted on antenna carrier

Two 20-dB standard gain horns have been mounted on an antenna carrier as depicted in fig. 3.6. The antenna carrier is mounted on the vehicle such that the antennas' H planes are parallel to the y-z-plane of the coordinate system fixed to the scene. The directivity patterns for 24 GHz in the u-v- and the v-w-planes have been determined numerically using MATLAB functions provided by Orfanidis [50]. Figures 3.7 and 3.8(a) show the directivity for $\beta = 0$ and $\alpha = 0$, respectively. The 3.5-dB beam-width is approximately $\pm 9°$ in α- and β-direction. The 3-dB width is almost equal, but looking at the -3.5-dB points allows an unambiguous comparison with the directivity of the patches discussed in the following subsection. As in SAR processing data obtained from different directions is integrated coherently, the phase of the antenna amplitude pattern plays an important role. Modifying the scripts that produced the power directivity patterns, the phase of the antenna amplitude pattern has been determined and is plotted in fig. 3.8(b). Within the beam-width as defined above, the phase varies by 14°, which might be negligible in many imaging applications. However, as α varies between 0° and 90°, the phase changes by approximately 450°, which for large coherent processing angle regions might cause severe image degradations unless compensated for. For a (point-like) target that is observable within exactly such a region of aspect angles, for which the phase of the two-way patterns varies by 360°, the processing result

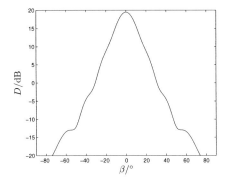

Figure 3.7: Directivity of a single horn antenna for $\alpha = 0$

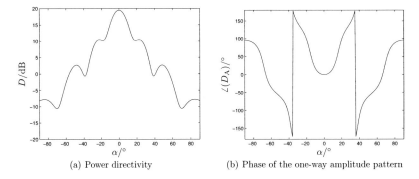

(a) Power directivity   (b) Phase of the one-way amplitude pattern

Figure 3.8: Directivity of a single horn antenna for $\beta = 0$

may be zero at the location of the target. This is a case very similar to the one, in which the phase of the pattern is constant, but the phase of the target's reflectivity varies by 360° within the region of observed aspect angles, as simulated in section 2.7.

Patch antennas

Patch antenna arrays for a center frequency of 24 GHz have been manufactured as transmit and receive devices that are arranged as shown in figures 3.9(a) and 3.9(b). TX and RX branches are implemented as arrays that narrow the beam with respect to β when the antennas are mounted such that the antennas' H planes are parallel to the y-z-plane of the coordinate system fixed to the scene. The patterns, determined by a simulation using Agilent ADS, are plotted for $\alpha = 0$ and $\beta = 0$ in fig. 3.10

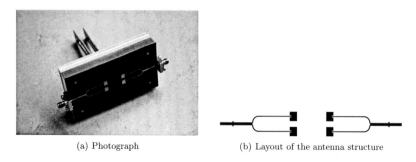

(a) Photograph   (b) Layout of the antenna structure

Figure 3.9: TX and RX patch arrays

and fig. 3.11, respectively. Each one of the TX and RX branches including the feed networks are symmetric with respect to the u-axis. However, each of them alone is not symmetric with respect to any line parallel to the w-axis. That means that the pattern of the transmit array—the same applies to that of the receive array—is expected to be symmetric in β-direction but not necessarily in α-direction, which is verified by the plots in fig. 3.10 and fig. 3.11.

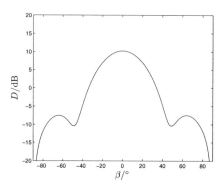

Figure 3.10: One-way directivity of patch array for $\alpha = 0$

According to the signal model developed in section 2.3, the signal received by the radar system is determined by the product of the patterns of the TX and the RX antennas, i.e. the received power is proportional to

$$D_{\text{equiv}}^2(\alpha) = D_{\text{TX}}(\alpha) \cdot D_{\text{RX}}(\alpha), \qquad (3.1)$$

where D_{equiv} is defined as an equivalent pattern, and the phase shift introduced by the

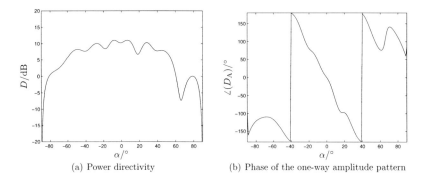

(a) Power directivity

(b) Phase of the one-way amplitude pattern

Figure 3.11: One-way directivity of patch array for $\beta = 0$

antennas is

$$2\angle(D_{A,\text{equiv}}(\alpha)) = \angle(D_{A,\text{TX}}(\alpha)) + \angle(D_{A,\text{RX}}(\alpha)). \tag{3.2}$$

For equal TX and RX patterns, the equivalent pattern is equal to each one of the patterns, as is the case for the horn antennas discussed above. For the antenna arrangement depicted in 3.9(a), the TX pattern is a flipped version of the RX pattern, since TX and RX array are symmetric about the v-w-plane. That means that the equivalent one-way pattern of the patch array arrangement is symmetric about $\alpha = 0$. Furthermore, the phase of the equivalent pattern varies to a much smaller extent than that of the horn antennas. This is due to the roughly linear phase variation of the amplitude pattern of the TX array that tends to be canceled by the phase variation of the mirrored pattern. Figure 3.12 shows plots of the equivalent TX-RX one-way pattern of the patch array

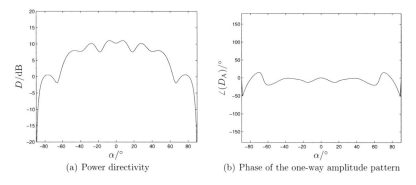

(a) Power directivity

(b) Phase of the one-way amplitude pattern

Figure 3.12: One-way directivity of TX-RX arrangement for $\beta = 0$

arrangement according to (3.1) and (3.2). The 3.5-dB beam-width is approximately $\pm51°$ in α-direction and $\pm21°$ in β-direction.

Performance comparison

The equivalent TX-RX one-way antenna patterns of the horn and patch arrangements discussed above have been investigated with respect to their $\mathrm{SNR_r}$ performance in section 2.9.3. There, however, each of the patterns separately has been compared to that optimum patterns, whose directivity, integrated over all solid angle elements from $\alpha = -90°$ to $\alpha = 90°$ for $\beta = 0$, has been equal to the integrated directivity of the horn and the patch pattern, respectively. That means that comparing the normalized results obtained there for the horn and the patch patterns is not meaningful. Directly comparing the horn and the patch patterns, and normalizing to the maximum value, yields the result shown in fig. 3.13. The plot reveals that maximum $\mathrm{SNR_r}$ is achievable

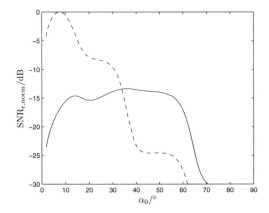

Figure 3.13: Normalized $\mathrm{SNR_r}$ achievable with horn antennas (dashed) and patch antennas (solid)

using the horn antennas and processing for aspect angles within $\pm7°$. For the same region of aspect angles, the $\mathrm{SNR_r}$ achievable with the patch antennas is 17 dB lower. As α_0 is increased, the $\mathrm{SNR_r}$ achievable with the horn antennas decreases, while that achievable with the patch antennas increases. For $\alpha_0 = 34°$, both antennas yield equal $\mathrm{SNR_r}$. For larger α_0, the $\mathrm{SNR_r}$ achievable with the patch antennas remains relatively constant up to approximately 60°, while the one achievable with the horn antennas drops rapidly. It becomes obvious that the antenna's beam-width alone is not sufficient to decide which one of two antennas is better suited for a given imaging task. As an example, consider data to be processed for α within $\pm20°$. Comparing the beam-widths of the horn antennas ($\pm9°$) and the patch antennas ($\pm51°$), one might conclude that the horn antenna's beam is too narrow and the patches are better suited. However, due to the horn antenna's higher gain, the resulting SNR_r is higher than that achievable using patch antennas even for integration angles larger than the horn antenna's nominal beam-width.

3.2.4 Position sensor unit

Figure 3.14: Position sensor unit

The signal received by the radar system has to be sampled equidistantly along the synthetic aperture. A rotary encoder, equipped with a metal disc carrying a rubber ring about its circumference, is used to determine the vehicle's position. The encoder delivers 4000 impulses per rotation. For the chosen circumference of the rubber ring, this yields one impulse per $41.405\,\mu$m. The position sensor unit (PSU), depicted in fig. 3.14, contains a PIC microcontroller (Microchip Inc.), which can be configured by the PC via RS-232. Every adjustable number of encoder impulses, the PSU issues triggers for the signal generator and the ADC. The PSU software ensures that—in case the vehicle is moved back and forth—a trigger corresponding to a certain position is issued only the first time this position is reached. Moving the vehicle back to positions already taken will not cause triggers to be issued. The next trigger will occur at the first trigger position that has not been taken yet. An LED display indicates the current position of the vehicle.

3.2.5 Analog-to-digital converter

A National Instruments DAQCard-6036E (PCMCIA version) is used as ADC. It offers a total sampling rate of 200 kS/s, resulting in 100 kS/s per channel when the inphase and the quadrature channel of the I/Q mixer are connected to the ADC, and a resolution of 16 bit over adjustable voltage ranges from $[-50\,\mathrm{mV}, 50\,\mathrm{mV}]$ to $[-10\,\mathrm{V}, 10\,\mathrm{V}]$. The ADC is configurable via the PC and delivers the acquired data to the PC. Data acquisition is triggered directly by the position sensor unit.

3.2.6 Personal computer

The PC is used to configure the signal generator, the position sensor unit and the analog-to-digital converter. Additionally, it stores the data acquired by the ADC and it can be used for SAR data processing using the software described in the following section.

3.3 Software

MATLAB software has been developed that allows to configure the SAR system and to manipulate and process the acquired radar data conveniently by offering graphical user interfaces (GUI). The key software components will be described in the following subsections.

3.3.1 Device configuration

Before operation, PSU, ADC and signal generator have to be configured. This can be accomplished using the device configuration GUI. In the following, the configurable parameters for each of the devices will be listed, along with useful information on the interaction of the devices.

Position sensor unit

The only parameter that needs to be configured regarding the PSU is the number of impulses, after which another trigger for the signal generator and the ADC is issued. On configuration of a new value, the LED position display is reset to zero. It is important not to move the radar system after having reset the PSU and before having started the ADC, since moving the system would cause triggers to be issued and consequently the signal generator to change the frequency. For the acquired data to be consistent, however, the first measurement has to be performed for the first frequency point rather than any other.

Signal source

The following parameters of the signal source can be configured:

- lowest frequency to be generated,
- frequency increment, and

- number of frequency points for one sweep.

From that parameters, a list of frequency points is generated and transmitted to the signal generator. After transmission, the signal generator waits for the next trigger event to happen. As soon as triggers, issued by the PSU, are received, the list of frequencies points is stepped through alternately in ascending and descending order.

Analog-to-digital converter

The following parameters of the ADC can be configured:

- sampling rate,

- sampling time per trigger, and

- number of triggers, for which data is to be acquired (corresponding to the length of the synthetic aperture).

After specifying a file name for the sampled data and optionally entering a comment, the ADC can be started. Upon each of the following trigger events, data is acquired and stored in the specified file in a format given by the MATLAB data acquisition toolbox.

3.3.2 Raw data manipulation

The file generated by the data acquisition toolbox contains the values measured for each of the trigger events corresponding to a certain position and a certain frequency. That data will be referred to as *raw* data. Depending on the configuration of sampling rate and sampling time per trigger, a certain number of values are sampled per trigger. Averaging those values belonging to the same trigger event results in a reduction of the influence of noise being present in the receiver. Since the ADC is triggered at the same time as the signal generator, the measured data might contain values influenced by transient effects within the settling time of the signal generator. The plots in fig. 3.15 shows inphase and quadrature channel raw data obtained in the absence of any targets in an anechoic chamber. A number of effects observable in the plots will be discussed in the following.

First, despite the absence of targets, the values are non-zero. This is due to the DC offsets caused by imperfect RF-to-LO isolation in the mixer MMICs and the non-zero coupling between TX and RX antennas. Since leakage and coupling depend on the signal frequency, the strength of the DC offsets varies with frequency.

Signal sections corresponding to a certain frequency are characterized by almost constant signal levels. The plots show data for 28 frequency steps each. The data have been acquired for 7 different frequencies, stepped through alternately in ascending and descending order. For each trigger, 100 samples have been taken per channel and plotted with one empty sample in between the sections belonging to one frequency. In each of the plots, the regularly ascending and descending order of the frequencies is

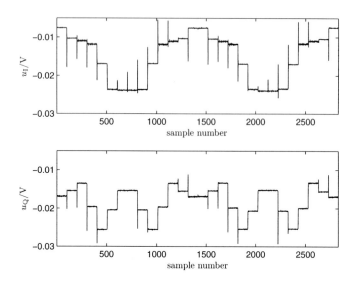

Figure 3.15: Raw data, obtained in the absence of targets

clearly visible. The average levels of samples belonging to triggers 8 to 14 are the same as those belonging to triggers 1 to 7 in reverse order. The average levels belonging to triggers 15 to 28 are a copy of those belonging to triggers 1 to 14.

The spikes following the trigger events are caused by transient effects that occur when the signal generator changes the frequency. The slight variations following the spikes are caused by noise. By means of the raw data manipulation GUI, the samples can be averaged per trigger. The user can specify which range of samples per trigger is to be considered and thereby exclude regions containing data subject to transient effects. Additionally, the user can specify the range of triggers to be considered, limiting the data to a certain region of the synthetic aperture.

3.3.3 Calibration

Data which have been obtained for a non-empty scene and which have been manipulated using the raw data manipulation GUI still contain DC offsets. Additionally, the mixers' gains might be different for different frequencies, and even different in I and Q channel for a given frequency. Not only the mixers' gains, but also the system's other components' transfer characteristics might vary with frequency. All those effects that have a multiplicative influence on the voltage that is sensed by the ADC, will be summarized in the term *gain*. In the following, two methods for both DC offset and gain calibration will be discussed.

Calibration by means of dedicated calibration measurements

A very obvious approach to canceling the DC offsets is performing a measurement in an anechoic chamber prior to the actual measurement of the scene, and subtracting the values measured for different frequencies from the values obtained for the same frequencies while passing the scene actually to be imaged. The remaining variations of the gains can be calibrated by using a reference target—in an anechoic chamber—in a known distance r_0 from the radar system. As—for a fixed arrangement of target and system—the frequency f_n is stepped through from f_{min} to $f_{min} + (N-1)f_\Delta$, where the nomenclature is the same as in section 2.5.3, the phase of the received signal is expected to vary as

$$2k_n r_0 = \frac{4\pi r_0}{c} \cdot f_n \tag{3.3}$$

according to (2.90) and the signal model developed in section 2.3, in case the target's reflectivity is assumed to be constant with varying frequency. That means that the I and Q signals are expected to vary harmonically with a period of $c/2r_0$ over the frequency. Comparing the values actually measured with those expected allows to determine the calibration factors for each of the frequency points. Figure 3.16 shows the I and Q values

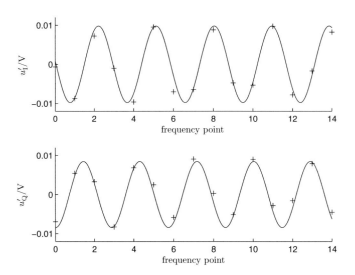

Figure 3.16: Measured (+) and expected (solid) I and Q signals after DC offset calibration

measured for a target in a certain distance, along with the expected values, which are harmonic functions of the frequency. The drawbacks of that calibration scheme become obvious from a look at those frequency points, for which the expected signal is close to zero, e.g. for frequency point 6 in the I channel and frequency point 5 in the Q channel. Considering that the expected signal might have been determined slightly erroneous with respect to its period or its phase, the determination of the calibration factor, which equals the ratio of measured and expected signal, is highly sensitive to a slight shift of the expected signal around one of its zeros. That means that the signal expected for a given target has to be known very exactly, and therefore, the target's distance has to be known to within fractions of a wavelength, which is on the order of 1.25 cm in the investigated system. Additionally, the calibration scheme does not work, when the expected signal is zero at one of the frequency points to be calibrated, since determination of the calibration factors involves a division by the expected signal.

The mentioned problems can be circumvented by performing a number of calibration measurements with varying target distances that are chosen such, that for each of the frequency points, at least one of the expected signals is non-zero, or preferably in the region of its maximum, where the sensitivity of the determination of the calibration factors with respect to an erroneous expected signal is minimum. Then, for each frequency point, the calibration factor can be determined by comparing received and expected signals for that target distance, for which the expected signal is maximum (compared to other target distances). The result of that scheme is a relative calibration of the gains for different frequencies for each of both channels separately. A potential gain imbalance of I and Q channel, resulting in phase errors as discussed in section 2.10.3, is then calibrated by adjusting the multiplication factors of both channels in such a way, that the mean powers in I and Q channel of the resulting calibrated data are equal.

The exact knowledge of the target's distances is rendered unnecessary by estimating the period of the expected signal from the signal actually received, rather than determining it as $c/2r_0$. Since according to (3.3) the signals in I and Q channel are expected to be harmonic, the period for each of the target's (unknown) distances can be estimated by investigating the spectrum of the received signals.

A GUI has been implemented that allows the user to calibrate imaging data according to the procedure described above. First, the data of a single frequency sweep for an empty scene (anechoic chamber) is necessary for DC offset calibration. Then, data obtained for calibration targets (e.g. metallic spheres) at varying distances in an anechoic chamber have to be supplied. The GUI allows to display, for each of the frequencies and both channels separately, the maximum value of the expected harmonic signals contained in the measurement data for different distances. Thus, the user can decide, whether—for each of the frequency points—the expected signal is far enough from zero, or whether additional measurements are necessary.

Still, there are drawbacks related to this method. First, an anechoic chamber is needed for calibration, and second, the calibration might become invalid as the system's operating point changes with e.g. time and temperature, demanding frequent calibrations. For a sufficiently large amount of radar data collected from the scene to be imaged, dedicated calibration measurements might be completely unnecessary. A method for system calibration based exclusively on the data obtained for the scene to be imaged will be described in the following.

Calibration by exploiting SAR signal properties

For a sufficient number of samples, the phase of the signal received in the presence of targets is a uniformly distributed random variable. Therefore, the expected values of the signals in I and Q channel are zero. The I and Q signals measured by the ADC are superpositions of the DC offsets and the signals due to target reflexions. That means, that the DC offsets—per frequency point and channel—can be removed by determining the mean values of the signals actually measured, and subtracting them from the measured signals.

Assuming that—for a random scene—the mean powers per frequency and channel, evaluated over the P positions taken by the system and indicated by p, have to be equal, i.e.

$$\frac{1}{P} \sum_{p=1}^{P} |s_{c,n,p}|^2 = \text{const} \quad \forall \quad c, n, \tag{3.4}$$

where c is the channel index and n the frequency index, the gain calibration factors are easily determined by choosing them such, that condition (3.4) is fulfilled for the calibrated data.

A GUI has been implemented that performs calibration according to that scheme, i.e. the mean powers per channel and frequency—after DC offset calibration—are determined, and for all channels and frequencies, the measured values are corrected by a factor given by the ratio of the mean power over all frequencies and channels and the mean power for the frequency and the channel under consideration.

Experiments have shown that the effort related to conducting dedicated calibration measurements not only is unnecessary, but that the quality of the SAR images generated from data calibrated according to the scheme discussed here is superior to that discussed previously.

The SAR images in fig. 3.17 clearly show a difference in the effectiveness of the calibration schemes. SFCW systems exhibit an unambiguous range given by (2.98). The images in fig. 3.17 have been generated for SAR data that have been recorded for a scene consisting of a single metallic sphere of radius 5 cm at a range of approximately 1.8 m for 15 frequency points, spaced by 80 MHz and centered at 24 GHz. The resulting unambiguous range is 1.875 m. That means that a target at a certain range produces a response in the image not only at its true location, but additionally at ranges being multiples of the unambiguous range before and after the actual position. One of the sources of the DC offsets is the coupling of TX and RX antenna, which appears as a target at multiples of the unambiguous range, e.g. at range 1.875 m for the given system parameters. It is the purpose of DC offset calibration to prevent this coupling from appearing as a target. Figure 3.17(a) shows the image generated from data calibrated using dedicated calibration measurements as described in section 3.3.3, fig. 3.17(b) the one obtained from data calibrated exploiting signal properties as described in this

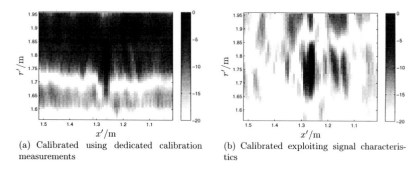

(a) Calibrated using dedicated calibration measurements

(b) Calibrated exploiting signal characteristics

Figure 3.17: SAR image for data calibrated according to different calibration schemes

section. As obvious from the images, the calibration scheme using only the actual measurement data outperforms the one that needs dedicated measurements in an anechoic chamber, since in fig. 3.17(a) the image of the metallic sphere is surrounded by a strong response due to the antenna coupling, whereas the level around the sphere's image in fig. 3.17(b) is much lower. A reason for this is the long-term instability of the DC offsets over time, which means calibration data is valid for a limited time span only. Using the properties of actual measurement data for calibration is far more effective and renders additional calibration measurements—to be performed in an anechoic chamber—unnecessary.

3.3.4 SAR processing

Another GUI has been implemented that allows the user to process calibrated data. The user has to specify the extent of the region to be reconstructed and recording parameters such as the used frequencies and sample spacing. Additionally, a system offset has to be specified that accounts for the length of the cables, which introduce a phase shift according to the wavelength inside the cables, but which do not cause the power density to decrease as for free-space propagation. For correct mapping of the reconstructed image to the actual scene and a correct compensation of the free-space loss according to the signal model developed in section 2.3, it is therefore necessary to consider the cables' length regarding a phase shift but not regarding free-space attenuation.

Furthermore, the user has to specify whether the antenna pattern is to be compensated and which additional weighting function in the x-domain is to be applied. Windowing functions for frequency weighting can be chosen. After specifying the integration limits $\pm \alpha_0$ and the desired number of sub-apertures, a SAR image is generated in the range-stacking like manner discussed in section 2.5.3, where the sub-apertures—if more than a single one is desired—are chosen to have equal angular extents, as in section 2.9.2.

The GUIs described above have been used to manipulate and process the data that have been acquired for discussion in the following chapter.

3 *System implementation*

4 Measurements

In the following sections, imaging experiments, conducted using the hardware described in the previous chapter, are documented. Scenes and processing parameters have been chosen to highlight a variety of characteristics of SAR imaging with an emphasis on short ranges and wide antenna beams. Processing has been done using the GUI described in section 3.3.4. The values displayed in the SAR images are $20 \log_{10} |\iota(x', r')|$. The x-axis of the plots will be displayed with increasing values from right to left according to the direction in which the aperture has been synthesized. As part of the processing, the influence of the used antennas is compensated—according to the patterns determined using analytical or simulated models in the previous chapter—prior to applying windowing functions. The sample spacings given for the following imaging experiments refer to the spacing between two samples taken for the frequency at the center of the signal bandwidth. Between those two samples, samples for the remaining frequency points are taken with accordingly smaller spacing in alternately ascending and descending order as described in section 3.2.1.

4.1 Cross-range and range resolution for non-zero signal bandwidth

In order to compare predicted and obtainable cross-range and range resolutions, the scene depicted in fig. 4.1 has been imaged. It consists of a single metallic sphere of radius $5\,\mathrm{cm}$ at a range of approximately $60\,\mathrm{cm}$ from the synthetic aperture, borne by a styrofoam pedestal and located within an anechoic chamber. As the radar system passes by the scene, it observes specular reflexions from that points of the sphere that are closest to the radar system. As long as those points, from which data are processed, are within a single resolution cell, the sphere may be considered a point-target. This is true for aspect angles up to $15°$ for the given setup when frequencies around $24\,\mathrm{GHz}$ are used. Data have been recorded using the horn antennas described previously. Predicted and actually obtained resolutions will be compared for $\alpha_0 = 5°$ and $\alpha_0 = 10°$. Cross-range resolution is mainly determined by the wavelength of the radiated signal and the angular extent of the observed region according to (2.65). For a frequency of $23.44\,\mathrm{GHz}$, the cross-range resolutions are predicted to be $\delta_x = 4.4\,\mathrm{cm}$ and $\delta_x = 2.2\,\mathrm{cm}$ for the integration limits $\alpha_0 = 5°$ and $\alpha_0 = 10°$, respectively. Range resolution for comparatively narrow antenna beams is determined by the bandwidth of the radiated signal. For a bandwidth of $560\,\mathrm{MHz}$, range resolution is predicted to be $\delta_r = 26.8\,\mathrm{cm}$ by (2.100). Figures 4.2(a) and 4.2(b) show SAR images generated from data acquired at 7 discrete frequency points spaced by $80\,\mathrm{MHz}$, starting from $23.44\,\mathrm{GHz}$ for the mentioned

4 Measurements

Figure 4.1: Setup for imaging a metallic sphere

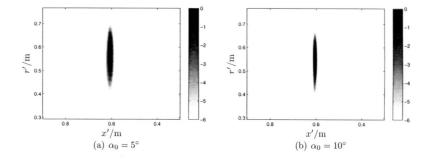

(a) $\alpha_0 = 5°$ (b) $\alpha_0 = 10°$

Figure 4.2: Resolution areas obtained for metallic sphere

values of α_0. The displayed values are normalized to the maximum value within the reconstructed region. The displayed dynamic range is 6 dB, which means that the gray area determines the achieved resolution. The extents of the gray areas are 4.0 cm and 2.2 cm in x-direction and 28.9 cm and 27.6 cm in r-direction for $\alpha_0 = 5°$ and $\alpha_0 = 10°$, respectively, which are in good agreement with the predicted values. Deviations may result from the fact that the sphere is not a real point-target, which will be discussed next.

Based on the simulation of imaging a metallic sphere with non-zero radius with a single frequency in section 2.7, one might consider the sphere's center as that position, at which the SAR image should exhibit its peak value. For a single frequency, the signal received along the synthetic aperture for concentric metallic spheres with zero

92

and non-zero radii are approximately equal in amplitude and differ in phase according to the ratio of wavelength and radius. Thus, the single-frequency SAR images of both spheres should exhibit one peak at a location that is equal in both images. The peaks should exhibit equal amplitudes, but their phase should differ by the amount given by wavelength and radius. When more than one single frequency is used, the peaks obtained for varying frequency are superimposed coherently. For the zero-radius sphere, the peaks' phases are equal, and therefore, the amplitudes will be superimposed in phase. For the non-zero radius sphere, the phases of the peaks obtained for each of the frequencies are unequal, which might result in destructive interference at that pixel of the image that corresponds to the sphere's center. Since the phase shifts are proportional to frequency and processing effectively causes the data per system position to be Fourier transformed (s. section 2.5.3), the peak is shifted towards the location of the sphere's surface. Therefore, the sphere used in the imaged scenario is an approximation to a point-target only.

4.2 Cross-range and range resolution for zero signal bandwidth

As discussed previously and shown by a simulation in section 2.7, a sphere is a rather good approximation to a point-target even if it exhibits non-zero radius, in case imaging is performed using one single frequency only. Resolution properties of single-frequency imaging have been validated by measurements using a metallic sphere of radius 4 cm at a range of approximately 75 cm as a target. Samples have been taken with a spacing $\Delta x = 3.1$ mm. For a frequency of 23.44 GHz, range resolution is predicted by (2.77) to be $\delta_r = 10$ cm and $\delta_r = 2.6$ cm for $\alpha_0 = 22.5°$ and $\alpha_0 = 45°$, respectively. Cross-range resolution is predicted by (2.65) to be $\delta_x = 1$ cm and $\delta_x = 5.4$ mm for the given values of α_0, respectively. The values obtained in reality can be estimated from the SAR images

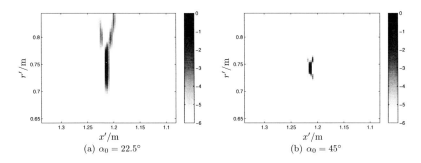

(a) $\alpha_0 = 22.5°$ (b) $\alpha_0 = 45°$

Figure 4.3: Resolution areas obtained for metallic sphere using a single frequency

in fig. 4.3 exhibiting a dynamic range of 6 dB and therefore showing the resolution areas, discretized with 3.1 mm in each direction. The extent of the resolution areas at the locations of the images' maximum values is on the order of 1 cm and 10 cm in x- and r-direction, respectively, for $\alpha_0 = 22.5°$, and 6 mm and 3 cm in x- and r-direction, respectively, for $\alpha_0 = 45°$, which are in good agreement with the theoretically predicted values.

4.3 Undersampled data

Using the patch antennas described previously, the received signals exhibit considerable magnitudes compared to those observed for $\alpha = 0$ even for large aspect angles. The resulting large spatial bandwidth causes the allowable sample spacing along the synthetic aperture to be comparatively small. Assuming a beam-width of $\pm51°$ in α-direction, the sample spacing has to be less than 3.9 mm for a maximum frequency of 24.56 GHz according to (2.103) in order to avoid aliasing. Figure 4.4 shows a SAR

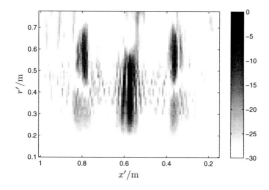

Figure 4.4: SAR image of a metallic sphere for undersampled synthetic aperture

image of a metallic sphere generated from data measured using 15 frequency points spaced by 80 MHz in the region from 23.44 GHz to 24.56 GHz, where the processing limits $\pm\alpha_0$ have been chosen to be $\pm10°$, and a Hann window has been applied with respect to the frequency points in order to suppress range side-lobes. Instead of the required spacing of 3.9 mm, it was 7.5 mm in the measurement. As expected, the SAR image shows strong responses in addition to the response at $x' = 58$ cm that corresponds to the location of the sphere at a range of approximately 40 cm. Under certain circumstances, a technique called digital spotlighting in [35] may help reduce aliasing. Alternatively, either the spatial sampling rate has to be increased, or an antenna with a narrower beam-width has to be chosen. A benefit of using antennas with comparatively wide beams, in conjunction with uniform angular weighting, will be shown in the next section.

Note that the horizontal bands that are visible in fig. 4.4 and will be visible in following SAR images are due to the fact that the processed signals are discretized with respect to x. The number of non-zero elements in the compensation vector depends on α_0, the range of the line to be reconstructed, and the sample spacing. For many of the adjacent range lines, the number of non-zero elements is equal. As the number of non-zero elements changes between two lines, an edge becomes visible, especially if the number of non-zero elements is small.

4.4 Angular weighting

In order to demonstrate the dependence of the image content on processing parameters, the target depicted in fig. 4.5 has been used. It is made of aluminum foil fixed on a sty-

Figure 4.5: Squinted target consisting of aluminum foil fixed on styrofoam

rofoam panel. The aluminum foil has horizontal and vertical extents of approximately 1 m and 30 cm, respectively. The foil is not completely planar, but it exhibits a certain roughness. The center of the foil is located in a distance of 1 m from the synthetic aperture. The foil is oriented such that its surface normal lies approximately within the u-v-plane of the antenna's coordinate system, and that it intersects the u-axis with an angle of approximately 10°.

Figure 4.6 shows SAR images generated from data obtained for 15 frequency points spaced by 80 MHz in the region from 23.44 GHz to 24.56 GHz using horn antennas and a Hann window with respect to the frequency points for range side-lobe suppression. Limiting the processing to $\pm\alpha_0 = \pm5°$, the SAR image (fig. 4.6(a)) consists mainly of two peaks originating from the scattering at the left and right edges of the foil. Since the foil is expected to act as a specular reflector for an aspect angle $\alpha = 10°$ due to its

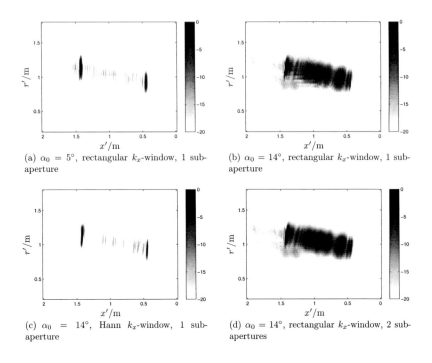

(a) $\alpha_0 = 5°$, rectangular k_x-window, 1 sub-aperture

(b) $\alpha_0 = 14°$, rectangular k_x-window, 1 sub-aperture

(c) $\alpha_0 = 14°$, Hann k_x-window, 1 sub-aperture

(d) $\alpha_0 = 14°$, rectangular k_x-window, 2 sub-apertures

Figure 4.6: SAR images of squinted aluminum foil, processed with varying parameters

orientation, the specular reflexions are outside the considered region which is limited by $\pm\alpha_0 = \pm 5°$. The non-zero values between the edges are caused by the roughness of the foil. Increasing the processing limits above $10°$, the specular reflexions become visible, as depicted for $\alpha_0 = 14°$ in fig. 4.6(b). The SAR image clearly reproduces position, extent and orientation of the target, but only for a proper weighting function. Using a k_x Hann window (fig. 4.6(c)) instead of a rectangular window (fig. 4.6(b)), the weighting introduced for the direction of the specular reflexion is so low that the image is comparable to the one obtained for $\alpha_0 = 5°$. The SAR image in fig. 4.6(d) has been generated using the same parameters as used for fig. 4.6(b), except that sub-aperture processing using 2 sub-apertures has been performed, resulting in a slightly smoother image, i.e. a reduction of speckle.

4.5 Roadside scenario

A scenario consisting of two cars and a concrete wall as depicted in fig. 4.7(a) has been imaged at 7 frequency points spaced by 80 MHz in the region from 23.44 GHz to 23.92 GHz using horn antennas. A Hann window with respect to the frequencies has been applied for range side-lobe suppression. The corresponding SAR image shown

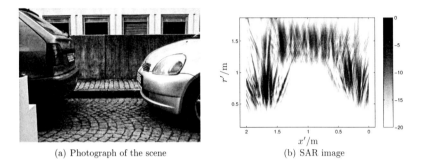

(a) Photograph of the scene   (b) SAR image

Figure 4.7: Roadside scenario: two cars and a concrete wall

in fig. 4.7(b) has been generated using 4 sub-apertures for $\alpha_0 = 20°$. The front and rear parts of the cars can vaguely be discerned. The gray area centered at a range of approximately 1.5 m does not originate from a target at that distance, but it is rather due to the concrete wall at a range of 3.4 m. The unambiguous range using a frequency spacing of 80 MHz is 1.9 m according to (2.98). Therefore, the concrete wall at a range of 3.4 m appears as a target also at range 1.5 m, amongst others. For the given scenario, the frequency spacing needs to be smaller in order to prevent the concrete wall from appearing as a target at undesired locations within the reconstructed area.

4.6 Bar code scenario

Optical bar codes are widely-used to identify objects in industrial environments. It might be desirable to use a similar concept at wavelengths that allow to read the codes even if the object is hidden behind a material being opaque for optical wavelengths. The scenario depicted in fig. 4.8(a) has been imaged using the patch antennas at one single frequency point (23.44 GHz) with a sample spacing $\Delta x = 3.1$ mm. The SAR image in fig. 4.8(b), which has been generated for $\alpha_0 = 20°$, shows that the aluminum patches, each with an area of approximately 1.5 cm^2, spaced by 15 cm in x-direction and located at a range of approximately 50 cm, are clearly discernable. This means that a short-range radar bar code system might be implemented at very low costs with a monofrequent system using inexpensive patch antennas. The metallic sphere shown in fig. 4.8(a), which is located outside the reconstructed area, ensures that radar

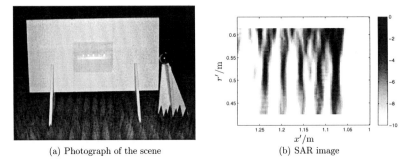

(a) Photograph of the scene (b) SAR image

Figure 4.8: Bar code scenario

data of sufficient strength and amount is acquired along the synthetic aperture so that the calibration scheme exploiting signal characteristics as described in section 3.3.3 is applicable.

4.7 Fawn scenario

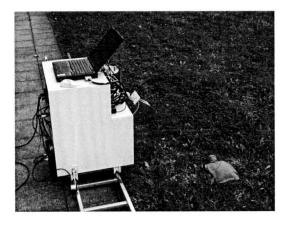

Figure 4.9: Fawn scenario

In [1], the problem of detecting wild animals during pasture mowing in order to prevent them from getting injured or even killed is addressed. As a model for fawns, a fabric-covered hot-water bottle made of rubber is proposed. Such a hot-water bottle in a meadow as depicted in fig. 4.9 has been imaged in two different configurations using horn antennas, directed towards the ground at a depression angle of approximately 45°. In the first configuration, 7 frequency points spaced by 80 MHz in the 560 MHz wide region from 23.44 GHz to 23.92 GHz have been taken, while in the second configuration 15 frequency points with the same spacing have been taken in the 1.2 GHz wide region from 23.44 GHz to 24.56 GHz. Sample spacing has been $\Delta x = 1.4$ mm and $\Delta x = 7.5$ mm, respectively. The measurements for both configurations have been conducted in different meadows, each with the hot-water bottle being located in a range of approximately 60 cm from the synthetic aperture. SAR images, generated for $\alpha_0 = 10°$ using a Hann window in x-direction, are shown in fig. 4.10. Due to the different band-

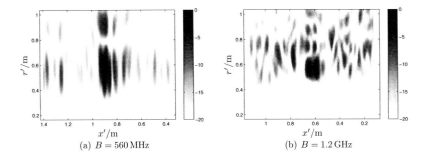

(a) $B = 560$ MHz　　　　　(b) $B = 1.2$ GHz

Figure 4.10: SAR images of fawn scenario

widths, range resolutions in fig. 4.10(a) and fig. 4.10(b) are different. In both images, the hot-water bottle, located approximately in the image centers, produces pixel magnitudes well in excess of the surrounding. This result suggests that SAR imaging, even using a low-cost system, might help in saving wild animals.

5 Conclusion

In this work, wide-beam, short-range synthetic aperture radar imaging is treated from the perspective of weighted mean reflectivity estimation.

The formation of SAR images relies on exploiting the fact that targets at different positions in a given scene contribute to the received signal with a characteristic phase signature. This phase signature is caused by the Doppler effect. However, SAR data can be recorded even when the system operates in a stop-and-go manner and does not move while radiating and receiving electromagnetic signals, i.e. the system works even if the received signal is not Doppler shifted with respect to the radiated signal. Section 2.2 clearly distinguishes between desired and undesired Doppler influences and clarifies the circumstances under which the undesired effects can be neglected.

In section 2.3, a signal model including phase and amplitude of the antenna directivity function is established. Ways to solve the inverse problem—finding the reflectivity distribution of the scene based on the observed signal—are discussed in section 2.4. Inverse filters—as opposed to frequently implemented matched filters—are considered as a starting point for further discussion since they yield estimates of the unweighted mean reflectivity of targets as observed from different aspect angles, while matched filters assign different weights to different aspect angles. Weighting contributions from different angles equally can be advantageous for scenes consisting of targets with unknown orientations. The influence of windowing functions used to control image properties like side-lobe levels on the angular weighting is discussed, as well as the way windowing functions have to be established in order to achieve the desired properties. Resolution issues are addressed, and the behavior of single-tone processing results in range direction is investigated, giving valuable insights for the implementation of very-low-cost systems.

The analytical investigations are complemented by simulations conducted for selected types of—partially hypothetic—targets. The resulting SAR images are references for the images obtained from real data discussed in chapter 4.

The values displayed in SAR images obtained from real-world data are not only corrupted by thermal noise influences, but also by the limited range of aspect angles from which targets are observable due to the width of the antenna beam. Those errors are referred to as caused by target noise. Figures of merit to judge the fidelity of SAR images regarding thermal noise and target noise are defined in section 2.8, and antenna patterns maximizing those figures of merit are derived. The orientation of the coordinate systems as chosen in section 2.1 allows the description of the relevant properties of the antenna pattern using a single coordinate rather than a combination of two. Optimum patterns for two extreme types of targets—omni-directional and strongly directive ones—are discussed for a variety of system and scenario parameters. Additionally, optimum patterns for weighted and sub-aperture processing are derived, and the influence of using sub-optimal patterns and various combinations of windows and antenna patterns are investigated. Phase errors caused by various system imperfections are discussed in section 2.10, and limits for those imperfections for given maximum errors are derived.

In chapter 3, mechanical, electrical and software components—and their interactions—of a 23- to 25-GHz SFCW wide-beam, short-range SAR system implemented with an emphasis on low costs are described. The price of the system's key components excluding analog-to-digital converters and signal processing hardware is below 150 €. The graphical user interfaces implemented allow the user to conveniently control the system and generate SAR images. Calibration, which is inevitable in a homodyne system due to the DC offsets present in the down-converted signal, is in many cases easily done exploiting characteristics of SAR signals—rendering dedicated calibration measurements unnecessary—and assisted by one of the graphical user interfaces.

In chapter 4, imaging experiments for a variety of scenarios using the system described in chapter 3 are documented. Among the considered scenarios are such that comprise isolated targets in an anechoic chamber and complex ones, like cars in a roadside scenario, a radar bar code scenario, and a fawn model in a meadow.

The presented theoretical investigations provide a variety of decision-making aids for the implementation of SAR systems targeted at certain imaging applications. The incorporation of antenna patterns and target characteristics into the formulation of imaging tasks has led to deep insights into the impacts of a variety of reconstruction paradigms on the information conveyed in SAR images. The benefits of uniform reflectivity weighting achieved by properly compensating the observed signals have been shown by means of numerical simulations and analytical derivations. The introduced figures of merit allow to judge the image quality to be expected for given system and scene parameters. Image quality for a system to be implemented can be maximized following the optimization procedures that have been derived. The SAR images, that have been generated using the described radar system, which operates around 24 GHz, and which has been implemented using dedicated radiofrequency hardware costing below 150 €, suggest the successful applicability of wide-beam, short-range SAR imaging techniques for such diverse applications as e.g. radar bar code reading and rescuing wild animals, even at very low costs.

Bibliography

[1] A. Patrovsky and E. M. Biebl, "Microwave sensors for detection of wild animals during pasture mowing," in *Advances in Radio Science, Kleinheubacher Berichte 2004*, vol. 3, 2005.

[2] C. Hülsmeyer, "Verfahren, um entfernte metallische Gegenstände mittels elektrischer Wellen einem Beobachter zu melden." German Patent no. 165546, 1904.

[3] W. Holpp, "The century of radar." http://100-jahre-radar.de/vortraege/Holpp-The_Century_of_Radar.pdf.

[4] C. A. Wiley, "Synthetic aperture radars," *IEEE Transactions on Aerospace and Electronic Systems*, vol. 21, no. 3, pp. 440–443, 1985.

[5] D. Felbach and J. Detlefsen, "Solving the inverse problem: a model based approach for circular cylindrical structures," in *IEEE International Geoscience and Remote Sensing Symposium*, vol. 1, pp. 29–31, July 2000.

[6] V. Shteinshleiger, V. Andrianov, A. Dzenkevich, V. Manakov, L. Melnikov, and G. Misezhnikov, "On a high-resolution space-borne VHF-band SAR for FOPEN and GPEN remote sensing of the Earth," in *IEEE Radar Conference*, pp. 209–212, 1999.

[7] A. Aguasca, A. Broquetas, J. J. Mallorqui, and X. Fabregas, "A solid state L to X-band flexible ground-based SAR system for continuous monitoring applications," in *IEEE International Geoscience and Remote Sensing Symposium*, vol. 2, pp. 757–760, 2004.

[8] J. C. Bennett, K. Morrison, A. M. Race, G. Cookmartin, and S. Quegan, "The UK NERC fully portable polarimetric ground-based synthetic aperture radar (GB-SAR)," in *IEEE International Geoscience and Remote Sensing Symposium*, vol. 5, pp. 2313–2315, 2000.

[9] N. Casagli, P. Farina, D. Leva, G. Nico, and D. Tarchi, "Ground-based SAR interferometry as a tool for landslide monitoring during emergencies," in *IEEE International Geoscience and Remote Sensing Symposium*, vol. 4, pp. 2924–2926, 2003.

[10] G. L. Charvat and L. C. Kempel, "Synthetic aperture radar imaging using a unique approach to frequency-modulated continuous-wave radar design," *IEEE Antennas and Propagation Magazine*, vol. 48, no. 1, pp. 171–177, 2006.

[11] B.-L. Cho, Y.-K. Kong, H.-G. Park, and Y.-S. Kim, "Automobile-based SAR/InSAR system for ground experiments," *IEEE Geoscience and Remote Sensing Letters*, vol. 3, no. 3, pp. 401–405, 2006.

[12] T. Hamasaki, L. Ferro-Famil, E. Pottier, and M. Sato, "Applications of polarimetric interferometric ground-based SAR (GB-SAR) system to environment monitoring and disaster prevention," in *European Radar Conference*, pp. 29–32, 2005.

[13] R. Herrmann and M. Sato, "Image reconstruction by a ground-based SAR system," in *IEEE International Geoscience and Remote Sensing Symposium*, vol. 6, pp. 4104–4107, 2005.

[14] J. H. Kim, M. Younis, and W. Wiesbeck, "Implementation of ground-based SAR demonstrator system for digital beam forming," in *IEEE International Geoscience and Remote Sensing Symposium*, vol. 6, pp. 4037–4040, 2005.

[15] J. J. Martinez-Madrid, J. R. Casar Corredera, and G. de Miguel-Vela, "High resolution imaging techniques in step-frequency subsurface radars," in *IEEE International Geoscience and Remote Sensing Symposium*, vol. 1, pp. 769–771, 1996.

[16] V. Mikhnev, Y. Maksimovitch, and P. Vainikainen, "Microwave reconstruction of underground targets using frequency domain measurements," in *European Microwave Conference*, pp. 1–4, 2000.

[17] M. Pieraccini, G. Luzi, D. Mecatti, L. Noferini, and C. Atzeni, "Ground-based SAR for short and long term monitoring of unstable slopes," in *European Radar Conference*, pp. 92–95, 2006.

[18] M. Pieraccini, G. Luzi, and C. Atzeni, "Terrain mapping by ground-based interferometric radar," in *IEEE International Geoscience and Remote Sensing Symposium*, vol. 39, pp. 2176–2181, 2001.

[19] L. Pipia, A. Aguasca, X. Fabregas, J. J. Mallorqui, C. Lopez-Martinez, and J. Marturia, "Mining induced subsidence monitoring in urban areas with a ground-based SAR," in *Urban Remote Sensing Joint Event*, pp. 1–5, 2007.

[20] H. Rudolf, D. Leva, D. Tarchi, and A. Sieber, "A mobile and versatile SAR system," in *IEEE International Geoscience and Remote Sensing Symposium*, vol. 1, pp. 592–594, July 1999.

[21] D. Tarchi, K. Lukin, D. Leva, J. Fortuni, A. Mogila, P. Vyplavin, and A. Sieber, "Implementation of noise radar technology in ground based SAR for short range applications," in *The Sixth International Kharkov Symposium on Physics and Engineering of Microwaves, Millimeter and Submillimeter Waves and Workshop on Terahertz Technologies*, vol. 1, pp. 442–444, 2007.

[22] Z.-S. Zhou, W.-M. Boerner, and M. Sato, "Development of a ground-based polarimetric broadband SAR system for noninvasive ground-truth validation in vegetation monitoring," *IEEE Transactions on Geoscience and Remote Sensing*, vol. 42, no. 9, pp. 1803–1810, 2004.

[23] F. T. Ulaby, R. K. Morre, and A. K. Fung, *Microwave Remote Sensing: Active and Passive*, vol. III. Norwood: Artech House, 1986.

[24] J. C. Curlander and R. N. McDonough, *Synthetic Aperture Radar*. New York: John Wiley & Sons, 1991.

[25] H. Klausing and W. Holpp, *Radar mit realer und synthetischer Apertur*. München: Oldenbourg, 2000.

[26] M. I. Skolnik, *Introduction to radar systems*. Boston: McGraw-Hill, 2001.

[27] C. A. Balanis, *Antenna Theory*. Hoboken: Wiley, 2005.

[28] M. Soumekh, *Fourier Array Imaging*. New Jersey: Prentice Hall, 1994.

[29] J. R. Bennett, I. G. Cumming, and R. A. Deane, "The digital processing of Seasat synthetic aperture radar data," *International Radar Conference*, pp. 168–175, 1980.

[30] C. Cafforio, C. Prati, and F. Rocca, "SAR data focusing using seismic migration techniques," *IEEE Transactions on Aerospace and Electronic Systems*, vol. 27, no. 2, pp. 194–207, 1991.

[31] R. K. Raney, H. Runge, R. Bamler, I. G. Cumming, and F. H. Wong, "Precision SAR processing using chirp scaling," *IEEE Transactions on Geoscience and Remote Sensing*, vol. 32, no. 4, pp. 786–799, 1994.

[32] R. Bamler, "A comparison of range-Doppler and wavenumber domain SAR focusing algorithms," *IEEE Transactions on Geoscience and Remote Sensing*, vol. 30, no. 4, pp. 706–713, 1992.

[33] J. M. Lopez-Sanchez and J. Fortuny-Guasch, "3-D radar imaging using range migration techniques," *IEEE Transactions on Antennas and Propagation*, vol. 48, no. 5, pp. 728–737, 2000.

[34] A. Potsis, A. Reigber, E. Alivizatos, A. Moreira, and N. K. Uzunoglou, "Comparison of chirp scaling and wavenumber domain algorithms for airborne low frequency SAR data processing," *Proceedings of SPIE*, vol. 4883, pp. 11–19, 2003.

[35] M. Soumekh, *Synthetic Aperture Radar Signal Processing with MATLAB Algorithms*. New York: Wiley, 1999.

[36] A. B. Carlson, *Communication Systems*. New York: McGraw-Hill, 1968.

[37] F. Gerbl and E. M. Biebl, "On the influence of windowing on the SNR maximizing antenna pattern for widebeam SAR techniques," in *ITG-Fachbericht International ITG-Conference on Antennas*, vol. 200, 2007.

[38] F. J. Harris, "On the use of windows for harmonic analysis with the discrete Fourier transform," *Proceedings of the IEEE*, vol. 66, January 1978.

[39] F. T. Ulaby, R. K. Morre, and A. K. Fung, *Microwave Remote Sensing: Active and Passive*, vol. II. Norwood: Artech House, 1986.

[40] D. L. Mensa, *High Resolution Radar Cross-Section Imaging*. Boston: Artech House, 1991.

[41] J. B. Pawley, *Handbook of biological confocal microscopy*. New York: Springer, 2006.

[42] L. M. H. Ulander and H. Hellsten, "A new formula for SAR spatial resolution," *AEÜ Int. J. Electron. Commun.*, vol. 50, no. 2, pp. 117–121, 1996.

[43] J. D. Murray, *Asymptotic Analysis*, vol. 48 of *Applied Mathematical Sciences*. New York: Springer, 1984.

[44] P. T. Gough and D. W. Hawkins, "Imaging algorithms for a strip-map synthetic aperture sonar: Minimizing the effects of aperture errors and aperture undersampling," *IEEE Journal of Oceanic Engineering*, vol. 22, no. 1, pp. 27–39, 1997.

[45] K. Iizuka, A. P. Freundorfer, K. H. Wu, H. Mori, H. Ogura, and V.-K. Nguyen, "Step-frequency radar," *J. Appl. Phys.*, vol. 56, pp. 2572–2583, November 1984.

[46] R. J. Beerends, *Fourier and Laplace Transforms*. New York: Cambridge Univ. Press, 2003.

[47] M. I. Skolnik, *Introduction to radar systems*. Boston: McGraw-Hill, 1990.

[48] F. Gerbl and E. M. Biebl, "SNR considerations for widebeam, short-range synthetic aperture radar processing," in *German Microwave Conference 2006*, March 2006.

[49] F. Gerbl and E. M. Biebl, "Balancing target noise against thermal noise—on the optimum beamwidth for mean radar reflectivity estimation," in *IEEE Microwave Symposium*, pp. 1213–1216, 2007.

[50] S. J. Orfanidis, "Electromagnetic waves and antennas." http://www.ece.rutgers.edu/~orfanidi/ewa/ch14.pdf.

[51] R. Ansorge and H. J. Oberle, *Mathematik für Ingenieure*, vol. 2. Berlin: Wiley, 2000.

[52] F. Gerbl and E. M. Biebl, "On the optimum antenna pattern for widebeam radar reflectivity estimation," in *Advances in Radio Science, Kleinheubacher Berichte 2006*, vol. 5, 2007.

[53] A. Moreira, "Improved multilook techniques applied to SAR and SCANSAR imagery," *IEEE Transactions on Geoscience and Remote Sensing*, vol. 29, no. 4, pp. 529–534, 1991.

[54] L. Råde and B. Westergren, *Springers Mathematische Formeln*. Berlin: Springer, 1997.

[55] J. C. Kirk, R. Lefevre, R. van Daalen Wetters, D. Woods, and B. Sullivan, "Signal based motion compensation (SBMC)," in *IEEE International Radar Conference*, pp. 463–468, 2000.

[56] J. C. Kirk, "Motion compensation for synthetic aperture radar," *IEEE Transactions on Aerospace and Electronic Systems*, vol. 11, no. 3, pp. 338–348, 1975.

[57] A. Moreira and Y. Huang, "Airborne SAR processing of highly squinted data using a chirp scaling approach with integrated motion compensation," *IEEE Transactions on Geoscience and Remote Sensing*, vol. 32, pp. 1029–1040, September 1994.

[58] A. Potsis, A. Reigber, J. Mittermayer, A. Moreira, and N. K. Uzunoglou, "Sub-aperture algorithm for motion compensation improvement in wide-beam SAR data processing," *Electronics Letters*, vol. 37, pp. 1405–1407, November 2001.

[59] Y. Lin, "A programmable 23- to 25-GHz frequency synthesizer with USB interface," Master's thesis, Technische Universität München, 2006.